FUNDAMENTALS OF
HYDRO- AND AEROMECHANICS

ENGINEERING SOCIETIES MONOGRAPHS

Fundamentals of
Hydro- and Aeromechanics

BASED ON LECTURES OF
L. Prandtl, Ph. D.

BY

O. G. Tietjens, Ph.D.

TRANSLATED BY
L. Rosenhead, Ph.D. (Cantab), Ph.D. (Leeds)

DOVER PUBLICATIONS, INC.

NEW YORK

Standard Book Number: 486-60374-1
Library of Congress Catalog Card Number: 57-4859

**Manufactured in the United States of America
Dover Publications, Inc.
180 Varick Street
New York, N.Y. 10014**

ENGINEERING SOCIETIES MONOGRAPHS

For many years those who have been interested in the publication of papers, articles, and books devoted to engineering topics have been impressed with the number of important technical manuscripts which have proved too extensive, on the one hand, for publication in the periodicals or proceedings of engineering societies or in other journals, and of too specialized a character, on the other hand, to justify ordinary commercial publication in book form.

No adequate funds or other means of publication have been provided in the engineering field for making these works available. In other branches of science, certain outlets for comparable treatises have been available, and besides, the presses of several universities have been able to take care of a considerable number of scholarly publications in the various branches of pure and applied science.

Experience has demonstrated the value of proper introduction and sponsorship for such books. To this end, four national engineering societies, the American Society of Civil Engineers, American Institute of Mining and Metallurgical Engineers, The American Society of Mechanical Engineers, and American Institute of Electrical Engineers, have made arrangements with the McGraw-Hill Book Company, Inc., for the production of a series of selected books adjudged to possess usefulness for engineers or industry but of limited possibilities of distribution without special introduction.

The series is to be known as "Engineering Societies Monographs." It will be produced under the editorial supervision of a Committee consisting of the Director of the Engineering Societies Library, Chairman, and two representatives appointed by each of the four societies named above.

Engineering Societies Library will share in any profits made from publishing the Monographs; but the main interest of the societies is service to their members and the public. With their aid the publisher is willing to adventure the production and dis-

tribution of selected books that would otherwise be commercially unpractical.

Engineering Societies Monographs will not be a series in the common use of that term. Physically the volumes will have similarity, but there will be no regular interval in publication, nor relation or continuity in subject matter. What books are printed and when will, by the nature of the enterprise, depend upon the manuscripts that are offered and the Committee's estimation of their usefulness. The aim is to make accessible to many users of engineering books information which otherwise would be long delayed in reaching more than a few in the wide domains of engineering.

ENGINEERING SOCIETIES MONOGRAPHS COMMITTEE

Harrison W. Craver, Chairman.

PREFACE

For about thirty years there has existed a certain trend to bring together again, as was the case in Euler's days, the theoretical or mathematical hydrodynamics and the so-called hydraulics which was based almost entirely on experiments.

During the brilliant development of theoretical hydrodynamics in the second half of the last century, contact with reality and with practical engineering problems was more and more lost. This was due to the fact that in this so-called classical hydrodynamics everything was sacrificed to logical construction and no results could be obtained unless they could be deduced from the basic equations. Yet, in order to overcome the mathematical difficulties, these equations were simplified in a manner which often was not permissible even as an approximation.

The hydraulics, on the other hand, which tried to answer the multitudinous problems of practice, disintegrated into a collection of unrelated problems. Each individual question was solved by assuming a formula containing some undetermined coefficients and then determining these by experiments. Each problem was treated as a separate case and there was lacking an underlying theory by which the various problems could be correlated.

From these remarks it can easily be concluded in what manner the subject is treated in the present book. It is this synthetic process between theoretical hydrodynamics and practical hydraulics which stands in the foreground. Theoretical considerations and deductions are not detached from the facts of experience but are put in proper relation to them wherever possible. Experimental results, on the other hand, are considered first of all with the object in view to bring the multitudinous facts of experience into relation with the fundamental laws and underlying theory.

The same arrangement used in the two volumes of the German edition has been retained with the exception that the derivation of the general Navier-Stokes equation is placed at the end of the present "Fundamentals," which corresponds to the first volume

of the German edition. The second volume, which is entitled "Applied Hydro- and Aeromechanics," is complete in itself. For its understanding one does not require familiarity with the contents of the "Fundamentals."

The notation used in the German version has been retained wherever possible. In particular, the vector notation employed, which is due to Gibbs, is the same as that in the original text.

The author wishes to acknowledge his indebtedness to Dr. J. P. Den Hartog for helping to prepare this edition.

O. G. TIETJENS.

SWARTHMORE, PA.,
January, 1934.

CONTENTS

CHAPTER IV

PART TWO

KINEMATICS OF LIQUIDS AND GASES

CHAPTER V

CHAPTER VI

CHAPTER VII

CHAPTER VIII

PART THREE

THE DYNAMICS OF NON-VISCOUS FLUIDS

CHAPTER IX

CHAPTER X

FOREWORD

When Dr. Tietjens, at that time my assistant at the Kaiser Wilhelm Institute for Flow Research, informed me of his intention to write a book on hydrodynamics and aerodynamics based upon his notes of my lectures, I gladly encouraged him to carry out this plan. I myself had intended for some time to publish these lectures in a suitably amplified form but had been unable to do this because of more urgent matters, having thus far not even begun the preliminaries to this work. I was therefore very much pleased that a book should appear in which one of my former students made available the contents of these lectures. For my own part, in order to assure the reader that the portions of the book which are based on my lectures are a faithful presentation of their subject matter, Dr. Tietjens and I agreed that I should look over those parts of the manuscript and, where necessary, bring them into agreement with my ideas as expressed in the lectures. This has been done, and I have also added in the second volume some new results which were included in my lectures after the departure of Dr. Tietjens. Moreover, Dr. Tietjens has added considerable material of his own, especially in the second volume, among which may be mentioned the description of experimental methods for pressure and velocity measurement, of equipment for measuring air resistance, and of the various methods for making flow lines visible. He also added some historical remarks and a discussion of the contemporary literature on the flow through pipes and channels of varying cross section.

For practical reasons, the book appears in two parts, so written that each can be used separately. The first volume contains the theory of the equilibrium of liquids and gases with practical applications to balloons and airships and then gives a comparatively rigorous treatment of motion of the "ideal" liquid with zero viscosity, and ends with a chapter on viscous fluids. The second volume is primarily devoted to technical applications and for the greater part employs methods which can easily be understood by practical engineers. However, in the treatment of

airfoil theory more complicated mathematical methods were unavoidable. The division of the subject into a theoretical and another more practical part has been influenced to a certain extent by the fact that there were two lecture courses on which the major part of the book has been based. The first course entitled "Hydro- and Aerodynamics" dealt in a more or less theoretical manner with the fundamentals of the science, whereas the second course "Aeromechanics" was more practical and in close touch with questions of aviation. However, the division of the subject into the two volumes of Tietjens' book does not strictly correspond to that of the two courses. The second volume, as was mentioned before, goes into several topics which were dealt with in my lectures and generally is much more elaborate. This is especially true of the exposition of airfoil theory, where the additional matter has been principally developed from publications of my own.

In connection with the fundamental tendency of my lectures, I wish to make some brief remarks. I have particularly endeavored to work out the fundamentals clearly. Details which can be found in other textbooks have been either entirely omitted or only briefly mentioned. In order to bring out the salient points I have, in presenting experimental results, relied more upon a few typical examples than upon many facts which can be found in the experimental literature. I therefore hope that on account of its individuality this book will obtain a place by the side of the other existing books on the theory of fluid motion, and I give it my best wishes on its way.

L. PRANDTL.

GÖTTINGEN,
May, 1929.

FUNDAMENTALS OF HYDRO-
AND AEROMECHANICS

INTRODUCTION

Fundamental Ideas.—Hydromechanics, of which aero-mechanics is a special branch, is concerned with the motion of fluids, which can be divided into three classes:

1. Viscous fluids,
2. Thin fluids,
3. Gaseous fluids.

There are, however, no sharp dividing lines between the three groups; no clear distinction can even be made between solid and viscous substances. Syrup is obviously a viscous fluid, but, on the other hand, pitch which customarily is assumed to be a solid shows in certain cases all the properties of a viscous fluid (even at room temperature); its characteristics depend upon the speed at which the deformations are produced. Pitch when broken up by a hammer behaves like a solid, the velocity of deformation being large; but if given sufficient time, it will leak out of a hole in the side of a barrel under the influence of its own weight so that it is a viscous liquid when the velocity of deformation is very small.

In fluids (the term includes both liquids and gases) the force necessary to produce a given deformation decreases with the velocity of deformation. Fluids differ in this respect from solids, in which the stress is proportional to the strain. In liquids and gases displacements of any magnitude can be produced by extremely small forces, provided the rate of displacement is sufficiently slow. Fluid bodies can be defined as those bodies in which the force necessary to produce a deformation approaches zero as the velocity of deformation is decreased.

The difference between viscous and thin liquids is best understood by imagining a solid body to be moved (1) in a viscous liquid like oil or syrup and (2) in a thin liquid like water, the density of the oil or syrup not being very different from that

of water. Viscous liquids oppose a much greater force to such a motion than thin liquids. The larger internal friction of viscous liquids causes a much greater resistance to deformation than that offered by water, alcohol, or ether, or similar liquids having little internal friction.

The thin-fluid and gaseous-fluid groups differ in their resistance to changes in volume; while liquids are incompressible at ordinary pressures, gases are easily compressed by small pressures and consequently increased in density. When, however, the pressure changes are so small that density variations can be neglected, thin liquids and gases obey the same laws. Thin liquids and gases, in so far as they are incompressible, are the province of hydromechanics and aeromechanics in the narrow sense of the words. Alterations of volume in a gas can be brought about in two ways. They can be due either to extremely high velocities (of the order of the velocity of sound) or to pressure differences such as are caused in the atmosphere by an increase of height of the order of a mile. The case of great velocities occurs among others in the flight of projectiles; this branch of science is known as "gas dynamics," whereas the pressure differences caused by height belong to the domain of dynamic meteorology.

Thus fluids of all kinds can be fitted into the following scheme:

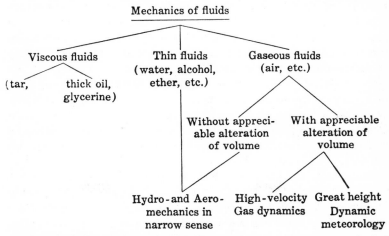

Mechanics of fluids

Viscous fluids
(tar, thick oil, glycerine)

Thin fluids (water, alcohol, ether, etc.)

Gaseous fluids (air, etc.)

Without appreciable alteration of volume

With appreciable alteration of volume

Hydro- and Aeromechanics in narrow sense

High-velocity Gas dynamics

Great height Dynamic meteorology

Historical Remarks.—A brief account of the history of the subject will make clear what is meant by the phrase "hydromechanics in the narrow sense of the word."

Although in classical and medieval times there was a certain amount of interest in hydrodynamics and more especially in hydrostatics, as is testified by the existence of the law of Archimedes, and although this knowledge was advanced by the work of Stevin, Galileo, and Newton, it is Leonhard Euler who is justly recognized as the father of hydromechanics. To him we owe our clear ideas on fluid pressure, and it was through his grasp of this fundamental concept that he later propounded the equations of motion bearing his name. He was a great theoretical mathematician, yet he brought his genius to bear on such technical matters as the construction of turbines.

But the great growth in technical achievement which began in the nineteenth century left scientific knowledge far behind. The multitudinous problems of practice could not be answered by the hydrodynamics of Euler; they could not even be discussed. This was chiefly because, starting from Euler's equations of motion, the science had become more and more a purely academic analysis of the hypothetical frictionless "ideal fluid." This theoretical development is associated with the names of Helmholtz, Kelvin, Lamb, and Rayleigh.

The analytical results obtained by means of this so-called "classical hydrodynamics" usually do not agree at all with the practical phenomena. To such extremely important questions as the magnitude of the pressure drop in pipes, or the resistance of a body moving through the fluid, theoretical hydrodynamics can only answer that both pressure drop and resistance are zero! Hydrodynamics thus has little significance for the engineer because of the great mathematical knowledge required for an understanding of it and the negligible possibility of applying its results. Therefore the engineers—such as Bernoulli, Hagen, Weissbach, Darcy, Bazin, and Boussinesq—put their trust in a mass of empirical data collectively known as the "science of hydraulics," a branch of knowledge which grew more and more unlike hydrodynamics.

While the methods of classical hydrodynamics were of a specifically analytical character, those of hydraulics were mostly synthetic. Hydrodynamics began with certain simple principles and, having made certain assumptions about the mechanical properties of fluids, attempted to formulate the behavior of the whole fluid mass from that of its elements. Hydraulics proceeded quite differently; it started out from the simple facts of experi-

ence and tried to explain complicated processes in terms of these. Hydrodynamics was built up on simplified natural laws, hydraulics on natural phenomena.

This difference in the outlook and method of the two sciences is connected with the difference in their purposes. In classical hydrodynamics everything was sacrificed to logical construction; hydraulics on the other hand treated each problem as a separate case and lacked an underlying theory by which the various problems could be correlated. Theoretical hydrodynamics seemed to lose all contact with reality; simplifying assumptions were made which were not permissible even as approximations. Hydraulics disintegrated into a collection of unrelated problems; each individual problem was solved by assuming a formula containing some undetermined coefficients and then determining these so as to fit the facts as well as possible. Hydraulics seemed to become more and more a science of coefficients.

Toward the end of the nineteenth and the beginning of the twentieth century a new critical spirit appeared in both sciences. The rapid growth of aeronautics and turbine engineering created a demand for more fundamental knowledge than "hydraulics" could provide. On the other hand, there grew up, among the physicists, a more realistic attitude associated especially with the great name of Felix Klein, having for its object a restoration of the unity between pure and applied science. Under these two influences, great progress was made in the synthesis of hydrodynamics and hydraulics.

It will be clear from these remarks about the evolution of hydrodynamics in what manner the subject will be treated here. The relation between theory and practice will always be kept in the foreground. Theory will not be detached from the facts of experience but will be put in its proper relation to them. Experimental results, on the other hand, will be considered in their relation to the fundamental laws and underlying theory.

PART I

THE STATICS OF LIQUIDS AND GASES

CHAPTER I

EQUILIBRIUM AND STABILITY

1. The Conditions under Which Liquids and Gases Can Be Treated as Continua.—We shall first consider the ways in which it is possible to describe the motions of a liquid or gas. A very general description of the motion of a fluid—considered as matter having a molecular structure—would be one in which the equations of motion were written down for each separate molecule. But apart from the fact that we know nothing about the intermolecular forces and that the mathematical difficulties are insuperable, we are generally not interested in the motions of the molecules or small elements of the fluid. The problem is rather to find out how the liquid behaves as a whole, or how such parts of it as contain a great many molecules behave; to find the velocity or acceleration or density or temperature at certain places. In other words it is not the motions of the molecules themselves which are of interest but rather their mean values in space and time.

We shall now discuss the way in which these mean values are obtained and point out those cases in which this cannot be done; that is to say, we shall point out those cases in which average quantities like "density" or "temperature" have no meaning. As an example we take the concept of density.

It is assumed that the liquid or gas is a system of discrete particles whose region of spatial distribution is V. The density of this system is defined as the quotient of the sum M of the masses of the particles in the volume V, divided by V, *i.e.*, the density in the volume V is M/V.

But how are we to define the density at a point? If A be the point inside the fluid at which the density is required, then we surround A by a series of small volumes ΔV_1, ΔV_2, . . . , so that each next volume is contained within the previous one ($\Delta V_{i+1} < \Delta V_i$). If we designate the sum of the masses contained in ΔV_i by ΔM_i, then we have in $\Delta M_i/\Delta V_i$ a series of difference quotients.

For relatively large values of ΔV the density is dependent upon the magnitude of ΔV (unless the gas or liquid is of uniform density throughout). As ΔV decreases, the variation in the density becomes less and less and finally the density appears to be reaching a limiting value (since it can be regarded as constant throughout sufficiently small volumes of the fluid). If, however, ΔV is made still smaller, fluctuations of ever increasing magnitude set in and, at the limit $\Delta V = 0$, $\Delta M/\Delta V$ is either zero or infinitely large. This is because, in the limit, A is either coincident with a particle or it is not. In the first case, $\Delta M/\Delta V$ has a definite numerator but a zero denominator, and thus the quotient is infinitely large; in the second case (where A is not coincident with a particle), the quotient $\Delta M/\Delta V$ will be zero and will remain zero in the limit.

The next question to be investigated is whether that value of ΔV for which $\Delta M/\Delta V$ remains constant can be regarded with sufficient accuracy as a "physical point"; whether therefore its volume is negligible in comparison with the dimensions of ordinary practice. Since air at normal temperature and pressure contains $2.7 \cdot 10^{19}$ molecules per cubic centimeter, it is clear that there can be no difficulties in this respect, for a volume of 10^{-12} cc (a cube of side $1/1,000$ mm) will contain $2.7 \cdot 10^{7}$ molecules, a number quite adequate for taking a mean value. In high-vacuum work, however, conditions are very different, since the mean free paths of the molecules may be of the order of magnitude of the dimensions of the apparatus, and the idea of density ceases to have any meaning. The curve shown in Fig. 1 gives the relation between the volume and density of a fluid. If we extrapolate this curve, as shown by the dotted line, through the point ΔV^0 to the left, we pass from the discontinuous to the continuous and from difference to differential quotients; or in mathematical language,

$$\lim_{\Delta V = 0} \frac{\Delta M}{\Delta V} = \frac{dM}{dV}.$$

The above discussion of the concept of density can be applied analogously to other mean values such as the kinetic energy of the molecules (which is a measure of the temperature). Therefore it is seen that in all cases in which there is a limiting value such as exemplified above, and in which the volume ΔV^0 can be regarded as a physical point, the molecular structure can be

disregarded and the system of particles can be replaced by a continuous medium. The accuracy of this procedure increases as the mean separation of the molecules decreases.

We have not yet taken into account the fact that the molecules are in a state of perpetual movement. If this molecular motion be disregarded, then each volume element will always contain the same molecules and will thus have a physical individuality; the effect of the molecular motion will be to destroy such physical individuality, since the molecules will be continually changing their position. If we make ΔV^0 so small that it is of the order of magnitude of the molecular mean free path, then the molecules will be continually passing in and out and it can no longer be

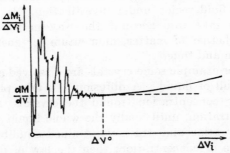

Fig. 1.—Connection between mean density and the volume from which it is derived.

said to have any physical individuality. Since, however, it is generally sufficient to work with volume elements large in comparison with the mean free path of the molecules, it is possible to regard such volume elements as having physical individualities for short periods and to neglect the constant interchange of molecules taking place at their boundaries. The fact that in long periods of time the molecular motion cannot be disregarded even in large volumes of fluid is shown by three phenomena: (1) in diffusion, in which there is a molecular interchange of matter; (2) in internal friction, in which there is an interchange of momentum between contiguous layers of fluid; and (3) in heat conduction, which is due to a molecular exchange of kinetic energy.

The continuum obtained by extrapolating the boundary value ΔV^0 to zero is an abstraction corresponding to a frictionless fluid in which infinitely small elements of volume retain their

physical individuality.   It is obvious that this mental abstraction is only a first approximation to a real fluid and is insufficient to investigate the laws of diffusion, friction, or heat conduction, or any process in which these phenomena are of importance. It is possible to use this simplified picture for viscous fluids provided they are in a state of equilibrium.   In general, however, it is necessary to introduce the molecular motion in one form or another in order to account for such phenomena as diffusion or heat conduction.   But since the volumes ΔV^0 contain a sufficient number of molecules to be regarded as physical entities for short periods of time, even viscous fluids can be treated as continua provided the mean free path of the molecules is absolutely negligible in comparison with the dimensions of the fluid region under investigation.   It is, of course, necessary to take into account the way in which molecular exchange (whether of matter, momentum, or energy) depends upon position and time.

Suppose for example some crystals are dissolved at the bottom of a vessel full of liquid.   A diffusion will take place from the region of big concentration (round the crystals) to regions of small concentration, until finally the whole liquid will have the same concentration and will have attained equilibrium.   Let n be the degree of concentration, then the law of diffusion (and similarly of viscosity or of heat conduction) can be derived if the fluid is assumed to be continuous while n is assumed to be a function of space and time.

There are cases, however, in which the differential equations have unstable functions as solutions; if it is desired to investigate the details of such an instability, it is necessary to make a critical reexamination of the problem, treating the fluid as a noncontinuum.   Such a reexamination is also necessary for stable, but rapidly varying, functions when the functional values (such as the density) are not constant even in the smallest volume ΔV^0 which can be treated as a physical entity.   Apart from these exceptions, however, we shall assume in our treatment that liquids and gases have molecular mean free paths of negligible dimensions and can therefore be regarded as continuous media. In order to emphasize that in exceptional cases liquids and gases differ in their behavior from continua, they are sometimes designated as "quasi-continua."

2. The Concept of Fluid Pressure.—The following may be taken as a kind of axiom on the subject which we are going to discuss:

"A continuum is in equilibrium when, and only when, the resultant of all the forces acting on any part of it is zero."

The forces acting can be divided into:

1. Surface forces, which are the forces acting on the surface of a body and are proportional to the superficial area.

2. Volume forces, mass forces, or body forces, which are proportional to the volume or mass. The attraction of the earth is an example of a volume or mass force.

In any body in equilibrium, on which external forces act, corresponding internal forces must be present (Fig. 2). It is easy to see that as the volume of the body is reduced to the limit and all forces acting on it are summed up, the internal forces disappear, since they can be separated into equal and opposite pairs. Thus we obtain as the condition necessary for the equilibrium of a continuum that the geometrical sum of all the external forces must be zero.

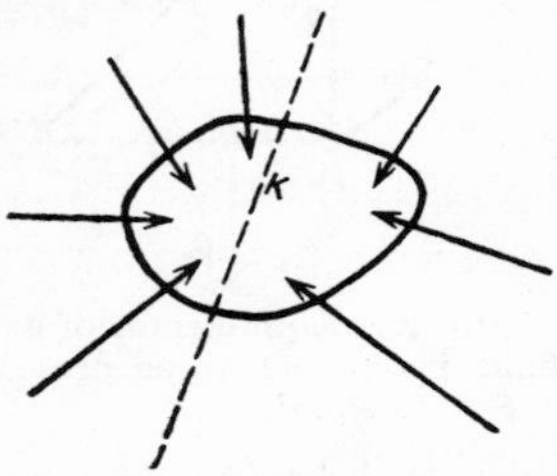

FIG. 2.—Body in equilibrium. The resultant of the forces over any surface K must be equal in magnitude and opposite in direction to the external forces acting on the body bounded by K.

We can obtain an insight into the internal forces at any part of the body by imagining the body to be cut along a plane K passing through the point under consideration, thus converting the internal forces into external ones. Since each half of the body is in equilibrium, the resultant of the internal forces over the cut surface must be equal in magnitude and opposite in direction to the resultant of the external forces acting on the cut body. Of course no information about the distribution of the internal forces over the cut surface is obtained in this manner.

A second characteristic (which may be treated either as a definition or as an experimental fact) of a fluid is:

In every part of a fluid in equilibrium the surface force is perpendicular to the surface on which it acts and is directed from the outside to the inside. This normality of the surface forces means nothing more than the complete absence of any frictional forces. The surface force per unit area is known as

"pressure" and, as was first shown by Euler,[1] this force has a particularly simple meaning for fluids in a state of equilibrium. By combining the experimental fact, or definition of a liquid, given above with the axiom enunciated at the beginning of this discussion, we shall now derive a fundamental conclusion about the pressure in a fluid. In doing so we shall make use for the first time of the statement contained in the axiom that it is applicable to a volume in the fluid of any form that we like to choose. To the volume thus chosen we apply the equilibrium condition that the surface forces are normal to the surface.

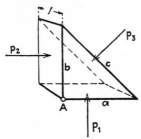

FIG. 3.—Equilibrium of a fluid prism; all three pressures are equal.

Consider then a volume of liquid in the form of a prism of unit height and with triangular end surfaces (Fig. 3). Let p_1, p_2, p_3 be the surface forces per unit area, *i.e.*, the pressures on the surfaces; for equilibrium the sum of all the vertical and of all the horizontal components must be zero. The forces normal to the end faces do not come into the problem since they have no component in the plane under consideration. Assuming volume forces to be absent, and representing the magnitude of the vectors p_1, p_2, p_3 by p_1, p_2, p_3, Fig. 3 shows that:

$$p_1a - p_3c \cos (a, c) = 0$$

and

$$p_2b - p_3c \cos (b, c) = 0.$$

But

$$a = c \cos (a, c)$$

and

$$b = c \cos (b, c),$$

so that

$$p_1 - p_3 = 0$$

and

$$p_2 - p_3 = 0,$$

and therefore

$$p_1 = p_2 = p_3. \tag{1}$$

This last equation does not depend upon the size of the prism or upon its orientation in space. If the volume of the prism converges to zero in such a way that the diagonal surface $c \times 1$

[1] EULER, L., General Principles on the Equilibrium of Fluids (French), "Histoire de l'académie," vol. XI, Berlin, 1755.

moves parallel to itself toward A, Eq. (1) can be interpreted as stating that in a fluid in equilibrium the force per unit area, or the pressure, at a point A is independent of the direction of the surface element on which it acts.

In other continua, such as elastic bodies or viscous liquids not in a state of equilibrium, the pressure at a point *does* depend on the direction of the plane on which it acts, and it can be described only by the form and orientation of an ellipsoid and therefore by six parameters. In a fluid in equilibrium, whether viscous or not, this ellipsoid degenerates into a sphere.

Now it is a fact of experience that the ability of a viscous fluid to take up tangential surface forces (shearing stresses) decreases with the viscosity. Thus taking the limiting case of zero viscosity we can ascribe to such an "ideal frictionless fluid" the property of being unable to resist shearing stresses.

In frictionless fluids the surface forces are always (not only in the state of equilibrium) normal to the surface of any element of volume in the fluid. The pressure at any point in an ideal frictionless fluid is therefore always independent of the orientation of the surface element on which it acts, that is to say the fluid pressure is uniquely determined by one parameter. Thus the pressure ellipsoid degenerates into a sphere in the case of frictionless fluids even when they are not

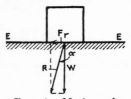

Fig. 4.—Motion of a rigid body of weight W along a plane E subjected to the action of a frictional force F_r. In equilibrium α is finite if the bodies in contact are rigid, but is zero if the bodies are fluid.

in a state of equilibrium. The pressure at a point in a frictionless fluid is therefore determined by a scalar, while the general state of pressure in a viscous fluid is characterized by a symmetrical tensor (*cf.* Art. 44). The resultant of the internal forces acting across a cut is always perpendicular to the plane of the cut in the case of a frictionless fluid, while in viscous fluids or elastic bodies this resultant is generally not perpendicular to the plane of the cut.

The fact that even for viscous fluids the pressure is normal to the surface of every fluid element in the case of equilibrium can be seen as follows. Consider a body of weight W which is moved along a plane E and subjected to a frictional force F_r (Fig. 4); the resultant of the forces exerted on the plane by the body will be inclined at an angle α to the normal of E. Now it is a property of fluids, in contradistinction to elastic solids,

that as the velocity of deformation decreases, α converges to zero. Thus, in equilibrium α is zero, so that even viscous fluids in equilibrium cannot sustain shearing forces.

We have neglected volume forces in deriving our ideas on fluid pressure, but we can easily generalize the above discussion to include cases where volume forces are present. Since surface forces are proportional to the second power and volume forces to the third power of the linear dimensions of any fluid element, the volume forces can be made negligible by making the fluid element small enough. But since in the derivation of the equation of fluid pressure the volume of the prism converged to zero, the above derivation is absolutely valid for the general case, in which both volume and surface forces are present.

3. Relation between Pressure Distribution and Volume Force.
We shall now consider how the pressure depends upon the volume

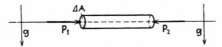

Fig. 5.—Equilibrium of a fluid cylinder whose generators are perpendicular to the direction of field of force; $p_1 = p_2$.

or body forces in a fluid in equilibrium. Let the unit of the gravity force **g** be the force acting on unit mass; this unit has the dimensions of an acceleration. Though the following treatment is applicable to any kind of body force, we shall discuss here only gravity since it is the most important body force in connection with aeronautical applications.

A field of force is characterized by a force vector **g** assumed to be a continuous function of position. Consider a narrow cylinder whose generator is perpendicular to **g**. For equilibrium in a fluid the resultants in three mutually perpendicular directions must be zero. Considering only the direction parallel to the axis of the cylinder it is clear that all the body forces and also the forces on the curved surface of the cylinder are perpendicular to the axis of the cylinder. We have, therefore, to deal only with the forces $\Delta A p_1$ and $\Delta A p_2$ (Fig. 5). These are in opposite directions and therefore the equilibrium condition is:

$$p_1 = p_2.$$

The position of the cylinder in the surface orthogonal to **g** was chosen quite arbitrarily. It is possible to arrange a great

number of such cylinders alongside each other such as to fill the whole surface perpendicular to **g** with them. Then, by making the cross-sectional area of the cylinders converge to zero, we can derive the following principle:

The pressure is constant in any surface orthogonal to the field of force.

These surfaces of equal pressure in the general case are known as "level surfaces," in analogy to the field of gravitation where the surfaces perpendicular to the lines of force are horizontal or level planes. If a field is not "surface-normal," that is to say has no continuous surfaces perpendicular to the forces, equilibrium cannot exist. A helical field for example has no orthogonal surfaces; a fluid mass under the action of such a field of force would begin to revolve and could not remain in a state of equilibrium.

A particularly important surface-normal field of force is the one whose vector **g** has a force function U, *i.e.*, whose vector is the gradient of a scalar $U = f(x, y, z)$:

$$\mathbf{g} = \text{grad } U,$$

or in coordinates

$$\mathbf{g}_x = \frac{\partial U}{\partial x}, \qquad \mathbf{g}_v = \frac{\partial U}{\partial y}, \qquad \mathbf{g}_z = \frac{\partial U}{\partial z}.$$

Such a field of force is called a "potential field." Since the surfaces of U = constant are orthogonal to the field of force **g** and are also, as shown above, surfaces of constant pressure, the pressure in a potential field is a function of U alone, *i.e.*,

$$p = f(U). \tag{2}$$

If the vector field **g** has no force function, it is in general also not surface normal; however, there are exceptional cases in which a vector field has no potential but still has normal surfaces. Such fields can always be brought into the form

$$\mathbf{g} = f(x, y, z) \text{ grad } U.$$

In general, however, equilibrium is then not possible, except with very special density distributions and even then the equilibrium is always unstable.

We shall now investigate the dependence of the pressure on the body force in the direction of the force vector. For this purpose we consider a cylinder whose generator is parallel to **g**

and whose end surfaces ΔA are coincident with two planes of constant pressure p_1 and $p_1 + dp$, respectively (see Fig. 6).

Applying the equilibrium condition in the direction of the force vector **g,** we see that forces on the curved surface of the cylinder do not enter into the problem since they are all perpendicular to the direction of the force component considered. Consideration of the remaining forces shows that the weight of the fluid cylinder is equal and opposite to the difference of the forces on the end surfaces of the cylinder.   If ρ is the mass per unit volume, then, using the symbols shown in Fig. 6, we have:

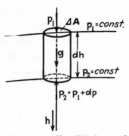

$$\rho \, dh \Delta A \cdot \mathbf{g} = (p_1 + dp)\Delta A - p_1 \Delta A = dp \Delta A$$

or

$$\frac{dp}{dh} = \rho \mathbf{g} \tag{3}$$

Thus we have derived an extremely important principle in the statics of fluids:

FIG. 6.—Equilibrium of a fluid cylinder whose generators are parallel to the field of force; $dp/dh = \rho\mathbf{g}$.

The pressure inside a fluid increases in the direction of body force in such a way that its increment per unit length is equal to the product of the density and the body force.

If we consider the increase in pressure in directions other than those perpendicular to $p = $ constant, it is obvious that dp/dh must be replaced by the gradient of p so that in general we can write

$$\text{grad } p = \rho \mathbf{g} \tag{3a}$$

or

$$\frac{\partial p}{\partial x} = \rho \mathbf{g}_x, \qquad \frac{\partial p}{\partial y} = \rho \mathbf{g}_y, \qquad \frac{\partial p}{\partial z} = \rho \mathbf{g}_z. \tag{3b}$$

For the case in which **g** is due to gravitational attraction, so that **g** is parallel to z, the equations simplify to

$$\frac{\partial p}{\partial x} = 0, \qquad \frac{\partial p}{\partial y} = 0, \qquad \frac{\partial p}{\partial z} = \rho g.$$

The sign of $\partial p/\partial z$ is correct if the positive axis of z is directed downward.   If the positive axis of z is upward, the formula is

$$\frac{\partial p}{\partial z} = -\rho g.$$

If the field of force has a potential, and if h is parallel to **g,**

$$\frac{\partial U}{\partial h} = \mathbf{g}. \tag{4}$$

Combining (4) with (3), we get

$$dp = \rho dU. \tag{5}$$

But since according to (2) the pressure is a function of U alone, we have

$$\frac{dp}{dU} = \rho = f'(U)$$

or

$$f(U) = \int \rho dU + \text{const.},$$

i.e., given the density distribution, $f(U)$ is simply determined by integration.

A result of the condition $\rho = f'(U)$ is that for $U = \text{constant}$ the density is also constant. Since we have already found that p is constant over surfaces of constant potential, we can say—with the aid of the equation of state of a gas $f(\rho, p, T) = 0$—that under the action of a potential force the state of the gas is constant at a level surface.

In many cases the dependence of the density upon the pressure is known, as for example in adiabatic changes in gases; we then have according to (5)

$$\int \frac{dp}{\rho} = U + \text{const.}$$

or

$$\int_{p_1}^{p_2} \frac{dp}{\rho} = U_2 - U_1,$$

where $\int \frac{dp}{\rho}$ can be calculated from a knowledge of the relation between p and ρ. The integral $\int \frac{dp}{\rho}$, which again is a function of p or ρ, is called the "pressure function" and is written

$$\int_{p_1}^{p_2} \frac{dp}{\rho} = P(p), \tag{6}$$

where p_1 is any arbitrary initial value of the pressure.

The pressure distribution in a fluid in equilibrium under the action of a potential force is completely determined in principle by this simple linear relation between the pressure function P and force function U.

4. Stable, Unstable, and Neutral Equilibrium.—An equilibrium is called stable if, after the application of an infinitely small displacement, the system returns to its original state; it

is said to be unstable if the displaced system tends to depart from its original state; and, finally, it is said to be neutral if the fluid can be given displacements of any magnitude without disturbing the equilibrium.

In order to find out the state of equilibrium, we imagine a small quantity of the fluid to be displaced through a suitable distance in a direction opposite to that of the force. If, after the displacement, it has a greater density than the surrounding fluid, it will obviously sink back again and tend to return to its former position. In this case there is stable equilibrium.[1]

If the displaced fluid has a smaller density than the surrounding fluid it will experience a hydrostatic lift and will therefore tend to move away from its original position; the liquid is then said to be in unstable equilibrium.

If a particle of fluid, after a small displacement in any direction, finds itself in a region of the same density as itself, the equilibrium is neutral.

If the density is independent of the pressure, *i.e.*, if the fluid is incompressible, and if differences in density are caused only by absence of uniformity in the fluid (as might be caused for example by different concentrations of salt in solution), then, if the direction of h is the same as that of the force, the fluid will be

in stable equilibrium, if $\dfrac{\partial \rho}{\partial h} > 0$;

in unstable equilibrium, if $\dfrac{\partial \rho}{\partial h} < 0$;

in neutral equilibrium, if $\dfrac{\partial \rho}{\partial h} = 0$.

In compressible fluids matters are somewhat more complicated since the various particles alter their density as well as their pressure with the position. In Chap. II, Art. 15, the stability of gases is discussed and the dependence of density upon pressure is then taken into account.

5. Equation of Hydrostatic Pressure.—As yet we have only assumed that the force vector **g** is a continuous function of the position and only in a few special cases the additional assumption was made that this vector field possesses a force function. We

---

[1] Of course a displacement in the direction of the force may also be imagined. For stability the displaced fluid element should then have a smaller density than its surroundings.

shall consider now in greater detail the special case where **g** is caused by gravity. The earth's attraction is a central force and has, like all central forces, a force function which we shall represent as before by U. If the curvature of the earth be neglected, U is constant in any plane parallel to the earth's surface. We take the direction of h as being positive from the earth's surface upward (opposite to our previous convention); there will therefore be a change in the sign of Eq. (4), so that

$$g = -\frac{\partial U}{\partial h},$$

or

$$g(h - h_1) = U_1 - U.$$

Since on the other hand from (5)

$$\int_{p_1}^{p} \frac{dp}{\rho} = U - U_1,$$

we have:

$$h - h_1 = \frac{1}{g} \int_{p}^{p_1} \frac{dp}{\rho}. \tag{7}$$

This gives a relation which makes it possible to carry out determinations of height by a measurement of the pressure and the temperature. From these two quantities the density can be calculated by means of the equation of state of the gas in question. This equation is particularly useful for the measurement of height in the air and is then known as the barometric-height formula.[1] The decrease in **g** with height due to the earth's spherical shape can usually be neglected, since at a height of 6 miles the error is less than $\frac{1}{5}$ per cent. For, if R is the radius of the earth and g_0 the acceleration of gravity on the surface, the acceleration g at a height h is given by

$$g = g_0 \left(\frac{R}{R + h} \right)^2 = g_0 \frac{1}{\left(1 + \dfrac{h}{R} \right)^2}.$$

Since $h \ll R$, this can be approximately written as

$$g = g_0 \left(1 - \frac{2h}{R} \right).$$

Therefore

$$-U = \int g\,dh = g_0 \int \left(1 - \frac{2h}{R} \right) dh$$

[1] We shall return to this equation and its application in Chap. II.

and

$$-U = g_0\left(h - \frac{h^2}{R}\right).$$

With the values

$$R = 4,000 \text{ miles and } h = 6 \text{ miles},$$

we find

$$\frac{h^2}{R} = 0.009 \text{ mile}.$$

This is the correction that must be applied to U and therefore to p in order to take into account the decrease in g due to the spherical shape of the earth.

It is convenient to write $\rho g = \gamma$, where γ is the weight per unit volume; then (7) becomes

$$h - h_1 = \int_p^{p_1}\frac{dp}{\gamma}. \tag{8}$$

The relation is particularly simple if ρ and therefore also γ can be assumed to be constant, as, for example, in the case of water at a uniform temperature. Equation (8) then becomes

$$h - h_1 = \frac{1}{\gamma}(p_1 - p)$$

or

$$p = p_1 - \gamma(h - h_1).$$

If we choose the origin of our system in such a way that $h_1 = 0$, *i.e.*, if we represent the pressure at the point $h = 0$ by p_1, then

$$p = p_1 - \gamma h, \tag{9}$$

or in differential form:

$$\frac{dp}{dh} = -\gamma, \tag{9a}$$

which is the *equation of hydrostatic pressure*. This important equation states in words that the pressure decreases in the upward direction, the decrement per unit length being equal to the weight per unit volume.

6. Application of the Pressure Equation; Communicating Vessels.—It is a direct result of the pressure equation that the liquid levels in two communicating vessels must lie in the same horizontal plane, because they are acted upon by the same atmospheric pressure. This result can also be obtained quite simply by applying the so-called *solidification* or *freezing prin-*

ciple, first used by Stevin.[1] Expressed in general terms this principle is:

A system in equilibrium remains so if any part of it is solidified subsequently.

Stevin, who applied this principle mostly to hydrostatic problems, pointed out that in a vessel with water at rest evidently any part of the water is in equilibrium and that this equilibrium cannot be disturbed if such parts of the water are supposed to be solidified, provided the density remains constant in the process.

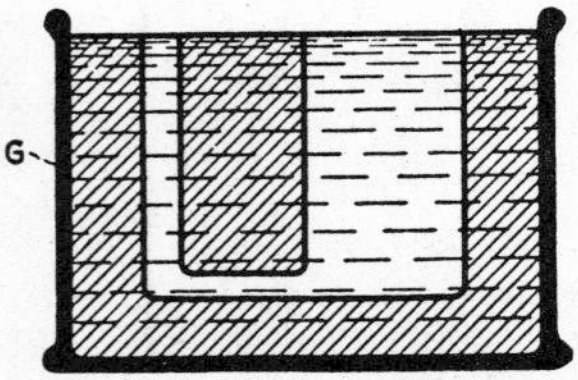

Fig. 7.—An open vessel filled with water. The law of pressure for open vessels in communication is obtained by imagining the portion shaded obliquely solidified without change of density.

In order to apply this principle to the problem of the communicating vessels, we imagine a vessel G to be filled with water (Fig. 7). We now freeze the water with the exception of two communicating vessels; the remaining liquid must then be in equilibrium in this form.

A practical application of such communicating tubes is found in liquid manometers which are frequently used in measuring pressures. If the manometer shown diagrammatically in Fig. 8a has a gas enclosed in its right stem and a water column of height h in the left stem, then

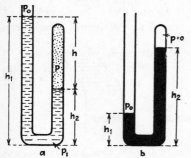

Fig. 8.—(a) Manometer for measuring gas pressure. (b) Manometer for measuring atmospheric pressure.

$$p_1 = p_0 + \gamma h_1$$
$$p_1 = p + \gamma h_2,$$

so that

$$p = p_0 + \gamma(h_1 - h_2) = p_0 + \gamma h,$$

where p is the pressure of the gas, p_0 the atmospheric pressure, p_1 the pressure in the manometer at its lowest part, and γ the weight per unit volume of the water.

The atmospheric pressure itself can be measured with this apparatus if there is a vacuum in the closed stem of the U-tube (see Fig. 8b); p is then zero and thus

$$p_0 = \gamma(h_2 - h_1).$$

[1] STEVIN, S., "The Principles of Hydrostatics" (Dutch), Leyden, 1586; also "Mathematical Papers" (French), Leyden, 1634.

Another example of the use of communicating vessels is found in the hydraulic press.   Figure 9 shows the principle of the press with which loads of almost any magnitude can be easily lifted. The two communicating cylinders of different bores are closed by well-fitting pistons, whose cross sections are A_1 and A_2

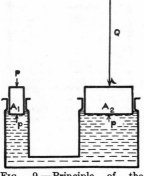

$(A_2 > A_1)$.   A force P on the smaller piston will produce a fluid pressure p given by

$$P = pA_1.$$

Owing to the pressure equality throughout the fluid the same pressure p will act upon the lower side of the piston of the greater cross section and produce the larger force

$$Q = pA_2.$$

Combining the two expressions:

FIG. 9.—Principle of the hydraulic press; $Q = \dfrac{A_2}{A_1} P$.

$$Q = P\frac{A_2}{A_1}.$$

7. Hydrostatic Pressure on Walls and Floors.—All the problems of this article will be solved by applying the hydrostatic pressure equation and remembering that the direction of the pressure is normal to the bounding surface.   The force

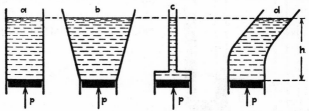

FIG. 10a–d.—Open vessels of different shapes; all have same pressure on base.

exerted by the fluid pressure on a horizontal surface (the bottom of a vessel) is

$$P = A(p - p_0) = A\gamma h.$$

The force on the bottom is therefore independent of the shape of the vessel and depends only upon the size of the bottom, the distance below the surface of the fluid, and the weight per unit volume of the fluid.   The vessels of various shapes shown in Fig. 10a to d all have the same pressure and the same force on the bottom.

The fact that very different quantities of fluid can produce the same force on the bottom, that the force exerted on the bottom of a vessel can in certain circumstances be greater than the weight of the whole fluid, is known as the "hydrostatic paradox" and can easily be explained by the principle of solidification. Imagine a cylindrical vessel with a bottom of area A to be filled to a height h with water; the force exerted by this liquid column on the bottom will be

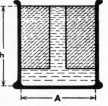

$$P = \gamma A h.$$

But the equilibrium and therefore the force on the bottom cannot be altered by solidifying part of the water (as shown cross-hatched in Fig. 11) so that the water remaining has the form shown in Fig. 10c. The other cases shown in Fig. 10 can be explained similarly.

Fig. 11.—Open vessel filled with water. Consider the portion shaded obliquely solidified without change of density; it is then evident that the pressure on the base is independent of the shape of the vessel, and dependent only on the height of the fluid.

The hydrostatic pressure on a wall can be calculated directly from the fundamental hydrostatic equation, but in this case the height h and therefore the pressure p are not constant over the surface. The force on a part of the surface of a vessel with perpendicular walls can be found by making use of the linear relation between pressure and depth. Construct a rectangular prism as shown in Fig. 12 and cut part of this prism off by a

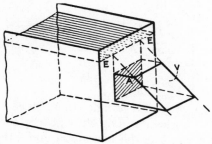

Fig. 12.—Hydrostatic pressure distribution on the wall of a tank.

45-deg plane, meeting the water surface at EE. Then the volume V of the cut-off part of the prism multiplied by the weight per unit volume of the fluid is equal to the force on the surface A. The height of the prism at any point of A evidently is equal to the pressure at that point.

Using the fundamental hydrostatic equation we obtain for the force P on the surface A:

$$P = \int^A (p - p_0)dA,$$

if p_0 is the pressure at $h = 0$ (atmospheric pressure); or since

$$p - p_0 = \gamma h$$

we find

$$P = \gamma \int^A h\,dA = \gamma A h_0,$$

where h_0 is the vertical distance between the center of gravity of the surface A and the liquid level.

We can obtain the point of application of the resulting force P by equating the couple of the force P to the resultant moment of all the individual pressure forces.

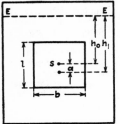

FIG. 13.—Point of application of side pressure on a rectangular surface.

$$Ph_1 = \int^A (p - p_0)dAh$$

or, with $p - p_0 = h\gamma$,

$$Ph_1 = \gamma \int^A h^2 dA = \gamma I,$$

if I is the moment of inertia of the area A.

If I is the moment of inertia about an axis through the center of gravity, we have

$$I = Ah_0^2 + I_0,$$

and, if we write (Fig. 13),

$$h_1 = h_0 + a,$$

then

$$Ph_0 + Pa = \gamma A h_0^2 + \gamma I_0 = Ph_0 + \gamma I_0$$

so that

$$a = \frac{I_0}{A h_0}.$$

If, for example, A is a rectangle, then, since the moment of inertia about the horizontal axis through the center of gravity is $I_0 = bl^3/12$ (where b is the horizontal and l the vertical side of the rectangle),

$$a = \frac{bl^3}{12blh_0} = \frac{l^2}{12h_0}.$$

If the upper end of the rectangle touches the water surface, then, since $h_0 = l/2$,

$$a = \frac{l}{6}.$$

8. Hydrostatic Lift and Stability.—The explanation of the fact that a body appears to lose weight when immersed in a fluid is easily obtained either from the solidification principle or with the help of the equation of hydrostatic pressure.

As we have seen, a fluid in a state of equilibrium will remain in equilibrium if any part of it is solidified without change of density. The weight of the solidified fluid mass is opposed by forces in the fluid which are greater on the under side than on the upper. The resultant of these forces must therefore be equal to the weight of the solidified mass; if the solidified mass be replaced by another body of the same shape, the forces on this body will be the same as before, *i.e.*, the lift will be the same.

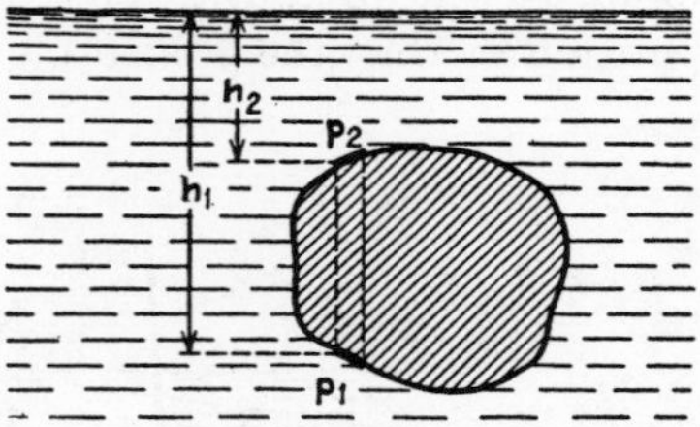

Mathematically the lift L is (Fig. 14)

$$L = \int^A (p_1 - p_2)dA$$

or, since $p_1 - p_2 = \gamma(h_2 - h_1)$,

$$L = \gamma \int^A (h_2 - h_1)dA$$

or

$$L = \gamma V,$$

where V is the volume of the

Fig. 14.—Hydrostatic lift. The resultant of the pressure forces over the whole surface is equal to the weight of displaced fluid.

body. Therefore a body completely immersed in a liquid experiences a lift, or apparent decrease in weight, of a magnitude equal to the weight of the fluid displaced and in such a way that the lift acts at the center of gravity of the displaced fluid. We thus have the principle:

The weight of a body floating on and partly immersed in a liquid is equal to the weight of the fluid displaced.

As regards stability it is clear that a body completely immersed in a liquid will be in stable equilibrium if its center of gravity is perpendicularly below that of the displaced fluid; it is in unstable equilibrium if its center of gravity is above that of the fluid, and the equilibrium is neutral if the two centers of gravity coincide. With bodies such as ships which are not completely immersed the conditions for stability are more complicated, since the form of the displaced liquid and consequently the position of its center of gravity are altered by the movement of the body.

Figure 15 shows the cross section of a ship diagrammatically; G is the center of gravity of the ship and B that of the displaced

fluid; B is therefore the point of application of the lift.   Suppose the ship is rotated about its center of gravity through a small angle $-d\varphi$, or, which is the same thing, suppose that the water line is rotated through $+d\varphi$ so that the center of gravity of the displaced fluid B moves to B'.   The weight of the body acting at G and the lift at B' are then of course referred to the displaced water line. Whether the ship is stable or not will depend upon the sign of the moment of these two forces set up by a displacement.   If we make the lines of action through B and B' intersect at M and denote the distance GM by h, the moment caused by the displacement will be $Whd\varphi$; for the conditions shown in Fig. 15 this moment will obviously tend to turn the ship back to its former position so that there is stable equilibrium.

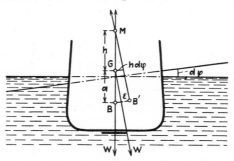

Fig. 15.—Section of a ship. Instead of rotating the ship through an angle $-d\varphi$ round the axis through the center of gravity G, the surface of the water is rotated through an angle $d\varphi$.

As can be seen from the geometrical construction, the equilibrium is stable as long as the point M is above G.   If M is below G, there is unstable equilibrium. M is called the "metacenter" and h the "metacentric height."   The higher the metacenter is above the center of gravity, the greater will be the stability.

9. Calculation of the Metacentric Height.—Let the weight of the ship, and therefore of the displaced water, be W and the point of application of the hydrostatic lift be B (Fig. 15).   Then a rotation of the ship through an angle $-d\varphi$ will shift the point of lift to B' through a distance ϵ.   The moment $W\epsilon$ must now obviously be equal to the algebraic sum of the moments due to the displaced water particles:

$$W\epsilon = \int^A \gamma dAyd\varphi y = \gamma d\varphi \int^A y^2 dA = \gamma I d\varphi,$$

where I is the moment of inertia of the water-line cross-section.

Thus

$$\epsilon = \gamma \frac{I}{W} d\varphi.$$

If we use the volume instead of the weight of the displaced fluid ($V = W/\gamma$), we have

$$\epsilon = \frac{I}{V} d\varphi.$$

From Fig. 15 it will be seen that

$$\epsilon = (h + a)d\varphi,$$

so that the metacentric height is given by

$$h = \frac{\epsilon}{d\varphi} - a$$

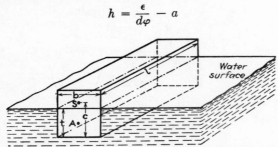

Fig. 16.—A prismatic body floating in water. Its metacentric height is
$$h = \frac{b^2}{12t} - a.$$

or

$$h = \frac{I}{V} - a. \tag{10}$$

The metacentric height can be determined experimentally by measuring the small angle $d\varphi$ produced by placing a large weight P first on one side of the ship and then on the other; the distance between the two sides may be denoted by b. From Fig. 15,

$$Pb = Whd\varphi$$

or

$$h = \frac{Pb}{Wd\varphi}.$$

We shall now determine the metacentric height of the body shown in Fig. 16, of length l, width b, and immersed depth t. Let the height of the center of gravity above the bottom surface be c. The moments of inertia of the water-line cross section are

$I_1 = lb^3/12$ for the longitudinal axis through the center of gravity, and

$I_2 = l^3b/12$ for the lateral axis through the center of gravity.

Equation (10) states that the stability decreases with the moment of inertia, and, since it was assumed that $l > b$ and therefore $I_2 > I_1$, the measure of the smallest occurring stability is

$$h = \frac{I_1}{V} - a = \frac{lb^3}{12V} - a.$$

Since

$$V = lbt$$

and

$$a = c - \frac{t}{2},$$

this becomes

$$h = \frac{b^2}{12t} - a = \frac{b^2}{12t} - \left(c - \frac{t}{2}\right).$$

So long as

$$\frac{b^2}{12t} > c - \frac{t}{2},$$

h will be positive and the equilibrium is stable. Therefore with constant depth and a constant height of the center of gravity from the bottom, the stability will depend markedly upon the width of the ship. At a certain width h is zero and the equilibrium is neutral; with further decrease in the width, h becomes negative and the equilibrium unstable.

CHAPTER II

APPLICATION OF THE PRESSURE EQUATION TO PERMANENT GASES. STABILITY OF AIR MASSES

10. Equation of State for Permanent Gases.—We start from the equation of state for permanent gases

$$pv = BT.$$

All the quantities in this equation are taken in engineering units: p in pounds per square foot, $v = 1/\gamma$ in cubic feet per pound, and T in degrees centigrade $+ 273$, so that B has the dimensions

$$[B] = \text{ft}/°\text{C}.$$

The magnitude of B generally varies with the different gases: For medium-damp air,

$$B = 96.5 \text{ ft}/°\text{C},$$

for dry air,

$$B = 96.1 \text{ ft}/°\text{C}.$$

Writing $\gamma = 1/v$ in Eq. (8) of Chap. I (page 20), we have

$$h_2 - h_1 = \int_{p_2}^{p_1} v \, dp, \qquad (1)$$

and eliminating v by the equation of state,

$$h_2 - h_1 = B \int_{p_2}^{p_1} T \frac{dp}{p}. \qquad (2)$$

It is seen that the temperature is here a primary factor.

To carry the calculation further it is necessary to make assumptions about the temperature distribution. We shall consider the following cases:

Uniform atmosphere,

$$v = \frac{1}{\gamma} = \text{const.}$$

Isothermal atmosphere,

$$T = \text{const.}$$

Polytropic atmosphere,

$$pv^n = \text{const.}$$

29

11. Uniform Atmosphere.—The assumption that the density of the atmosphere is independent of the pressure and therefore of the height is chiefly of value only for approximation formulas for small differences in height. The actual state of the atmosphere does not agree at all with this assumption, since it entails unstable equilibrium as will be seen later.

Under the supposition $v = 1/\gamma =$ constant,[1] we have from (1)

$$h_2 - h_1 = v \int_{p_2}^{p_1} dp = v(p_1 - p_2). \tag{3}$$

Taking the earth's surface as our level of zero height ($h_1 = 0$), and letting the constant density be equal to that at the earth's surface, $\gamma_1 = 1/v_1$, the result (3) becomes

$$h = v_1(p_1 - p), \tag{4}$$

The height h at which the pressure is zero will be the height of the uniform atmosphere. Let this height be h_0, then

$$h_0 = p_1 v_1, \tag{5}$$

and from the equation of state,

$$h_0 = BT_1,$$

where T_1 is the absolute temperature at the surface. The height of the uniform atmosphere is an important quantity. If we write $h_\star$ for the value of this height at a surface temperature $t = 0°C$ the normal height of the uniform atmosphere equals

$$h_\star = BT_0 = 96.5 \times 273 = 26,300 \text{ ft. or about 5 miles.} \tag{6}$$

In order to find the height of the uniform atmosphere for a surface temperature $t°C$, we can put

$$h_0 = h_\star \frac{273 + t}{273} = h_\star \left(1 + \frac{t}{273}\right),$$

or approximately

$$h_0 = h_\star(1 + 0.004t).$$

In conclusion we shall find at what height the pressure decreases by 1 per cent, *i.e.*, at what height $p_1 - p = p_1/100$. Equation (4) states:

$$h = v_1(p_1 - p) = \frac{v_1 p_1}{100}.$$

---

[1] The assumption that $v =$ constant is to a large extent true with liquids, so that in agreement with reality we can write

$$h_2 - h_1 = \frac{p_1 - p_2}{\gamma} \text{ (equation of hydrostatic pressure)}.$$

With the Eqs. (5) and (6) this becomes

$$h = \frac{h_0}{100} = 260 \text{ ft.}$$

Thus in a uniform atmosphere at a surface temperature of 0°C the pressure decreases every 260 ft by 1 per cent.

12. Isothermal Atmosphere.—We shall now suppose that the temperature is constant throughout the whole atmosphere:

$$T = \text{const.} = T_1.$$

Then, from Eq. (2),

$$h_2 - h_1 = BT_1 \int_{p_2}^{p_1} \frac{dp}{p}$$

or

$$h_2 - h_1 = BT_1 \ln \frac{p_1}{p_2},$$

and, from (5),

$$h_2 - h_1 = h_0 \ln \frac{p_1}{p_2}$$

or with $h_2 - h_1 = h$

$$p_2 = p_1 e^{-\frac{h}{BT_1}} = p_1 e^{-\frac{h}{h_0}}. \tag{7}$$

When p_2 is set equal to zero in this formula, it is seen that the height of the isothermal atmosphere is infinite. The equation also states that the pressure decreases exponentially with the height. Since the barometer reading b is proportional to the pressure we can also write Eq. (7) in the form

$$h_2 - h_1 = h_0 \ln \frac{b_1}{b_2} = h_\star \left(1 + \frac{t}{273}\right) \ln \frac{b_1}{b_2}.$$

(Barometric height formula)

Emden[1] has introduced the quantity $b_1/b_2 = n$ and drawn up tables of

$$H(n) = h_\star \ln n.$$

The whole discussion holds however only for $T = $ constant. In reality there usually is a decrease of temperature with increasing height, especially at great heights. In applying the barometric formula, therefore, it is customary to imagine the air to be divided into layers of constant temperature and to assign a mean temperature value to each separate layer.

[1] EMDEN, R., "Principles of Balloon Operation" (German), Leipzig, 1910.

13. Polytropic Atmosphere.—The polytropic atmosphere is defined by the equation:

$$pv^n = \text{const.}$$

where n may have any value; for $n = 1$ we have the special case of an isothermal atmosphere.

Eliminating v by the equation of state $pv = BT$, we obtain

$$p\left(\frac{BT}{p}\right)^n = \text{const.}$$

or

$$p^{\frac{1}{n}}\frac{BT}{p} = \text{const.}$$

or

$$T = \frac{\text{const.}}{B}p^{\frac{n-1}{n}}.$$

If $T = T_1$ when $p = p_1$, then

$$T = T_1\left(\frac{p}{p_1}\right)^{\frac{n-1}{n}}. \tag{8}$$

Introducing the expression for T given by (2), we get

$$h_2 - h_1 = BT_1 p_1^{-\frac{n-1}{n}}\int_{p_2}^{p_1} p^{-\frac{1}{n}}dp$$

or

$$h_2 - h_1 = h_0\frac{n}{n-1}\left(\frac{p}{p_1}\right)^{\frac{n-1}{n}}\Bigg|_{p_2}^{p_1}$$

or

$$h_2 - h_1 = \frac{n}{n-1}h_0\left\{1 - \left(\frac{p_2}{p_1}\right)^{\frac{n-1}{n}}\right\}. \tag{9}$$

This formula gives the relation between pressure and height.

An analogous connection between temperature and height can be obtained by introducing

$$\left(\frac{p_2}{p_1}\right)^{\frac{n-1}{n}} = \frac{T_2}{T_1}$$

into Eq. (9). This gives

$$h_2 - h_1 = \frac{n}{n-1}h_0\left(1 - \frac{T_2}{T_1}\right)$$

or

$$h_2 - h_1 = \frac{n}{n-1}B(T_1 - T_2), \tag{10}$$

stating that in a polytropic atmosphere the temperature is a linear function of the height.

If we again take the surface of the earth as the origin of our height scale by setting $h_1 = 0$, and instead of h_2 write the variable height h, Eqs. (9) and (10) become

$$p = p_1\left(1 - \frac{n-1}{n}\,\frac{h}{h_0}\right)^{\frac{n}{n-1}}$$

and

$$T = T_1\left(1 - \frac{n-1}{n}\,\frac{h}{h_0}\right). \tag{11}$$

The temperature is seen to vary linearly with the height, but the pressure-height diagram is a parabola of order $\frac{n}{n-1}$. For $p = 0$, *i.e.*, at the vertex of the parabola, a height $h = h'$ results from the relation

$$\frac{n-1}{n}\frac{h'}{h_0} = 1$$

or

$$h' = \frac{n}{n-1}h_0.$$

If $n = 1.2$, which is approximately the value observed with a semihumid adiabatic atmosphere, the height of the atmosphere is

$$h' = 6h_0 = 30 \text{ miles.}$$

In reality the atmosphere can be separated into two different regions. The lower part of it is called the "troposphere" and at our latitude is about 7 miles high. It is in this region that there is a continuous movement composed of large and small circulations. The large circulations are due to air streaming from the poles to the equator and back, the small circulations are governed by high- and low-pressure regions and limited in extent. The troposphere increases in height at the equator to about 9 miles, while at the poles it is only 4 miles. In the troposphere n can be given a mean value of 1.2.

Above the troposphere is the second part of the atmosphere, called "stratosphere," which is a stratified region in a state of radiation equilibrium without wind movement. It is a region of almost constant temperature, $t = -50°C$; this corresponds to the isothermal exponent $n = 1$. There is no sharp boundary between the troposphere and the stratosphere; they rather blend gradually into one another.

14. Determination of the Exponent n of the Polytrope.—The value of n can be found either from the relation between the temperature and the pressure or from that between the height and the temperature.

From (8) one derives

$$\frac{T_2}{T_1} = \left(\frac{p_2}{p_1}\right)^{\frac{n-1}{n}}$$

or

$$\log \frac{T_2}{T_1} = \frac{n-1}{n} \log \frac{p_2}{p_1},$$

and solved for n,

$$n = \frac{\log \dfrac{p_2}{p_1}}{\log \dfrac{p_2}{p_1} - \log \dfrac{T_2}{T_1}}.$$

It is permissible to substitute barometric heights for pressures in this formula.

The second method of finding n results from the relation between height and temperature as follows: Eq. (10) on rearrangement gives

$$n = \frac{h_2 - h_1}{h_2 - h_1 - B(T_1 - T_2)};$$

dividing by $h_2 - h_1$,

$$n = \frac{1}{1 - B\dfrac{T_1 - T_2}{h_2 - h_1}} = \frac{1}{1 + B\dfrac{T_2 - T_1}{h_2 - h_1}},$$

or in differential form

$$n = \frac{1}{1 + B\dfrac{dT}{dh}}, \tag{12}$$

or, solved for dT/dh,

$$\frac{dT}{dh} = -\frac{n-1}{n}\frac{1}{B}.$$

It is usual in meteorology to deal with the change in temperature for a change in height of 100 ft, which makes the temperature gradient—which we shall call Δ— of order of magnitude of $0.3°C/100$ ft. Thus:

$$\Delta = 100\left(-\frac{dT}{dh}\right).$$

Substituting the temperature gradient Δ in Eq. (12) and taking $B = 96.5$, we get

$$n = \frac{1}{1 - \dfrac{B}{100}\Delta} = \frac{1}{1 - 0.965\Delta} = \frac{1.04}{1.04 - \Delta} \qquad (13)$$

or solved for Δ,

$$\Delta = 1.04\frac{n - 1}{n}. \qquad (13a)$$

n can be determined from this equation when Δ has been found by experiment.

We have yet to consider an important special case. In general we are concerned with large masses of air which are unable to give up or receive heat in any rapid movement (except when the movement is one involving a mixing or radiation process) so that they are in an adiabatic state and obey the law

$$pv^k = \text{const.},$$

where $k = c_p/c_v = 1.405$. In our terminology (since the adiabatic is a special case of the polytrope)

$$n = 1.405.$$

On substitution of this value into Eq. (13a), it is found that the adiabatic temperature gradient is

$$\Delta = 0.30°\text{C}/100 \text{ ft.}$$

15. Significance of the Temperature Gradient in Relation to the Stability of Air Masses.—The adiabatic arrangement of air masses brings us to the new and important problem of the stability or instability of the atmosphere. We shall see that the adiabatic arrangement corresponds to a state of neutral equilibrium. By a stable arrangement we mean one in which a quantity of air, when brought from one height to another, is forced back to its original position again; if, on the other hand, the displaced air tends to move farther away from its original position, the equilibrium is said to be unstable. A neutral equilibrium is one in which a quantity of air can be displaced, without showing a tendency to return to its original position or to move farther from it.

In order to find out whether a mass of air is in stable or unstable equilibrium it may be imagined to be moved from one plane to a higher one. In doing so the air will expand adiabatically.

The problem is now to determine whether the air surrounding the displaced mass has a smaller, equal, or greater density than the displaced mass after its adiabatic expansion. Smaller density will mean stable equilibrium, equal density will mean neutral equilibrium, and greater density will mean unstable equilibrium. It is clear that an adiabatic arrangement is in neutral equilibrium, for every quantity of air moved from one position to another will find itself in a region of the same density as itself.

Therefore it is not a sufficient condition for the stability of a mass of air that the upper regions have a smaller density than the lower; the decrease in density must be at least that corresponding to the adiabatic law. We shall now investigate the part played by the temperature gradient in the different states of equilibrium. It will be seen that the temperature gradient is a convenient criterion of the stability or instability of a mass of air.

As we have seen, adiabatic conditions correspond to neutral equilibrium. In this case the temperature gradient is $\Delta = 0.30°C/100$ ft with $n = k = 1.405$. Since $n < k$ corresponds to stability, and $n < k$ means $\Delta < 0.30°C/100$ ft $(13a)$, we have stable equilibrium when

$$\Delta < 0.30°C/100 \text{ ft.}$$

Since, moreover, $n > k$ corresponds to instability,

$$\Delta > 0.30°C/100 \text{ ft.}$$

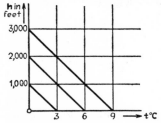

Fig. 17.—Relation between temperature and height for different surface temperatures. We assume an adiabatic air distribution; ($\Delta = 0.3°C/100$ ft).

means that the equilibrium is unstable.

Thus a temperature gradient of about $0.3°C/100$ ft is the greatest that can be observed in a stable air region, because larger gradients mean instability. Taking

$$\Delta = 0.30°C/100 \text{ ft}$$

as the approximate adiabatic temperature gradient, and plotting the height h as a function of the temperature t, we obtain a series of straight lines corresponding to different surface temperatures. In Fig. 17 a length along the abscissa corresponding to 3°C has been made equal to a length representing 1,000 ft along the ordinate, and the curves therefore cut the axes at 45 deg. For each surface temperature there is such a curve corresponding to an adiabatic distribution of air.

As we have seen, a temperature gradient of 0.3°C/100 ft, or a straight line inclined at 45 deg. to the axes, represents an air mass on the boundary between stability and instability. If Δ is smaller, *i.e.*, if the curve is steeper, the air distribution is definitely stable; smaller slopes correspond to instability.

Observations show that Δ can even be negative so that the temperature increases with height; this means a particularly stable distribution. Meteorologists call such a distribution an "inversion" (Fig. 18). Inversions can generally be explained as boundaries between air currents of different origin. The warmer layer of air pushes itself over the colder layer (the inverse case would be unstable).

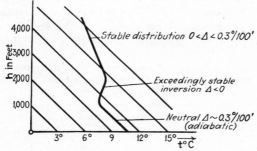

Fig. 18.—Air distribution involving inversion.

16. Influence of Humidity.—The air is in general mixed with water vapor. At any given temperature there is a definite limit to the amount of water vapor, that 1 cu ft can contain. As soon as this limit (saturation) is passed, part of the water is condensed and separated out in the form of tiny drops, which together form a cloud.

The quantity of water vapor at saturation per cubic foot of air is dependent upon the temperature but independent of the air pressure. The following table gives these saturation quantities in pounds per 1,000 cu ft at a number of temperatures:

Degrees centigrade:	−10°	0°	10°	20°	30°
Pounds per 1,000 cu ft:	0.12	0.30	0.59	1.08	1.90

As long as the water is in the gaseous state, its presence in small quantities makes little difference in the behavior of the air as regards its stability. If water condenses, however, it affects the behavior of the air considerably, because of the great heat of

condensation thereby set free (about 1080 B.t.u./lb. at 20°C). Condensation can occur for example if damp air is expanded adiabatically. Unsaturated air is soon saturated by adiabatic cooling since an adiabatic expansion produces a fall of temperature which brings down the saturation limit more rapidly than the volume of air expands. Condensation occurs as soon as the limit is passed, and a part of the water condenses on any dust particles present. It is to be noted that further adiabatic expansion will produce a slower rate of temperature decrease because of the heat of condensation liberated.

The temperature gradient in damp air (100 per cent relative humidity) under adiabatic conditions can be calculated, and for the surface layer it is found to be about half the value of the gradient in dry air. At higher regions the difference between dry and damp air is small on account of the smaller water content of air at lower temperatures, as shown in the following table:

VALUES OF THE POLYTROPIC EXPONENT n AND OF THE TEMPERATURE GRADIENT Δ (IN DEGREES CENTIGRADE PER 100 FT) FOR THE SATURATED ADIABATIC

Height, ft	Earth's surface temperature			
	$t = 0°C$		$t = 20°C$	
	Δ	n	Δ	n
0	0.19	1.22	0.13	1.15
6,800	0.23	1.28	0.15	1.17
13,600	0.27	1.35	0.17	1.20
20,400	....	....	0.19	1.23

As an example of a saturated adiabatic, we shall consider the following process. A quantity of saturated air is raised from some position to a higher one (from A to B, Fig. 19); such an increase in height can be produced when an air stream meets a range of mountains. Condensation will take place and will continue as long as the air is raised, and in certain cases the cloud drops may become sufficiently large to fall to the earth in the form of rain. If the cloud comes down on the other side of the mountain into a valley, the temperature of the air will at first rise according to the damp adiabatic curve, since part of the heat will be used up in vaporizing what is left of the cloud previously

formed.   From the moment C, Fig. 19, when the last drop of moisture has been vaporized, the rise of temperature will follow the dry adiabatic curve.   By the time the air has returned to its original height it will have experienced a considerable rise in temperature.   The process here described is the explanation of the so-called "Föhn," a hot wind blowing down from the Alps. In this manner large mountain ranges can influence the weather considerably.   Rainy regions always have uprising damp air, while the regions behind a mountain range where the air comes

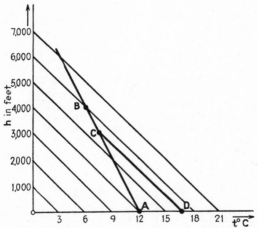

Fig. 19.—Example of air motion with "saturated adiabatic." The air mass rises from A to B (due to mountains) with increasing condensation; it falls from B to C with evaporation taking place—and then falls from C to D according to the law of the "dry adiabatic."

down with decreased moisture are rain free and dry.   Thus the west coast of South America is extremely rainy and fertile, because the west wind (damp from having blown over the ocean) there meets the Andes mountains, while the land behind is desert-like.

17. Concept of Potential Temperature.—The temperature which a given quantity of air would attain when moved under adiabatic conditions to some standard height, such as sea level, or when subjected to the pressure at that standard height, is known as the "potential temperature."

As we have seen in Art. 15, a temperature gradient $\Delta \lessgtr 0.3$ corresponds to a stable/unstable equilibrium so that curve I

in Fig. 20 represents an unstable, curve II a stable, and curve III a neutral equilibrium distribution in the atmosphere.

If the scale of abscissas is changed from ordinary temperature to potential temperature, the curves of Fig. 20 transform into

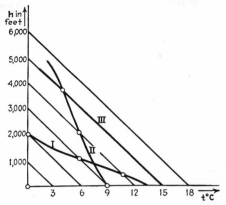

Fig. 20.—Curve I. Unstable distribution. Curve II. Stable distribution. Curve III. Neutral distribution.

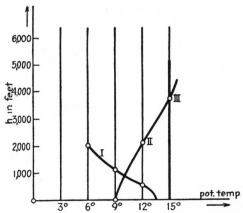

Fig. 21.—These are the curves corresponding to those of Fig. 20, when the abscissa is "potential temperature."

those of Fig. 21. Thus it can be said that a quantity of air (unsaturated with water vapor) is in stable/unstable equilibrium if the potential temperature increases/decreases upward (Fig. 21).

The potential temperature therefore plays the same part as the ordinary temperature in a constant volume of liquid with a positive temperature coefficient of expansion (for example

water above 4°C). If h be plotted as a function of the potential temperature, the adiabatic lines corresponding to various surface temperatures will be parallel to the h-axis and such lines mean a state of neutral equilibrium. If the potential tempera-

ture increases or decreases with increasing height, the distribution is stable or unstable respectively. The tangent to a stable curve at the point where it cuts the abscissa makes an acute angle α with the abscissa; the corresponding tangent to an unstable curve makes an obtuse angle β. The smaller α, or the greater β, the greater will be the stability or instability respectively (Fig. 22). The transformation of a

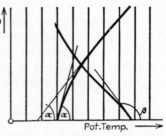

Fig. 22.—Stable and unstable air distribution according as the angle at the abscissa is smaller or greater than 90°.

height-temperature curve into a height-potential temperature curve is shown in Fig. 23a and b.

We shall now discuss the process of the mixing of stable layers of air. The thick black line in Fig. 24 represents a stable

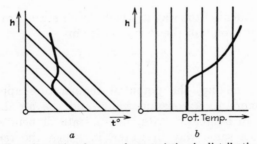

Fig. 23a–b.—Transformation from a characteristic air distribution (h, t) curve (a) to a (h, pot. temp.) diagram (b).

air distribution. If a mass A_1 of air at height h_1 is to be mixed with another mass A_2 at height h_2, both masses must be brought to the same height. Let us bring the mass A_1 from h_1 to h_2. The temperature adiabatically increases from t_1 to t_1' during the process. The mass of air A_2 at height h_2 and temperature t_2 now mixes with the mass of air A_1 at temperature t_1' and the two masses form a mixture with a temperature t_3. Conversely if the mass A_2 of air be displaced from h_2 to h_1, its temperature will become t_2' and the resultant mixture will have a

temperature t_4.   It is seen that, as a result of the mixing process, the temperature gradient approximates to the adiabatic one.

Since the temperature of the air masses at the region of mixture (h_2) differs from the potential temperature only by an additive quantity dependent upon the pressure, and since the potential temperature of the mass A_1 (displaced to h_2) is the same as at h_1 (because the displacement is adiabatic), the behavior of air masses on mixing can be determined by the use of the potential temperatures of the air masses.   This rule is, to be sure, only approximately correct, since the pressure distribution

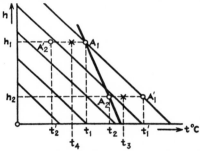

Fig. 24.—Effect on temperature gradient of mixing stable air distributions.

is not independent of the way in which the air strata are arranged. But the error is only of the second order in

$$\frac{h_1 - h_2}{h_0}.$$

By repeated mixing, the potential temperature approximates more and more to a constant value and the air distribution approaches an adiabatic state, *i.e.*, a state of neutral equilibrium.   In the same way in water, in which the temperature increases upward, repeated mixing will equalize the temperature which tends toward a constant value.

18. Origin of Clouds.—Changes of potential temperature in a stratified mass of air are due mostly to the action of the earth's surface and also to the effects of condensation and evaporation.   Other factors are the emission or absorption of radiation by the dust in the air and, less frequently, an emission from the air itself. Radiation on the air is chiefly absorbed by water vapor and carbon dioxide.

A cooling of the surface of the earth always produces a very stable arrangement (an inversion).   Figure 25 shows the process

of cooling as it takes place in the evening or night. It will be seen that as the cooling of the surface continues, higher and higher strata in the air are affected (chiefly by long-wave radiation absorbed by water vapor). We have here an example of increasing temperature with increasing height, a state known as an inversion and, as we have seen, such a state is particularly stable. In mountainous districts the valleys are often filled up in the evening with cold air which streams down from the slopes and then flows farther down the valley (valley wind).

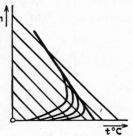

FIG. 25.—Surface cooling, as a result of which a very stable distribution is produced. (Evening and night.)

A rise in temperature at the surface of the earth due to sunshine produces an unstable distribution; rising currents of air and corresponding down currents are the result (Fig. 26). The inversion of the night (A_0 to A_n) is thereby smoothed out and the heat boundary A_1A_2 moves continually higher. Each rising air current goes a little higher than the previous one and then spreads out where it reaches equilibrium. The air between these ascending currents sinks downward to fill up the gaps. The whole of the layer in neutral equilibrium is

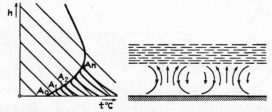

FIG. 26.—Surface warming, as a result of which an unstable distribution is produced. (Morning.)

continually stirred up. The thermal expansion of the lower air layers lifts the upper layers somewhat. This displacement has not been shown in Fig. 26 for the sake of simplicity.

At a sufficient height the rising air currents reach a region of condensation, *i.e.*, mists or clouds are formed; and thereafter the distribution curve is a saturated adiabatic. With uniform humidity and uniform initial temperature, condensation takes place simultaneously for all the parallel air currents and the

clouds have a horizontal bottom appearing to be floating on a horizontal mass of air (Fig. 27). We can therefore say that clouds are rising air currents in which condensation takes place as the result of oversaturation, since, as was mentioned previously, the air always contains a certain quantity of water vapor. The relative humidity continually increases with increasing height, *i.e.*, with decreasing temperature, until at a

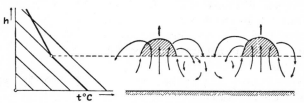

Fig. 27.—Formation of clouds as a result of condensation in rising air streams.

certain height saturation is attained and condensation occurs. Up to this height the rising air currents are invisible; above it they appear as clouds.

If the upper layers have sufficiently small stability, the clouds experience a lifting force; this will happen if the temperature gradient in the layers above is greater than the temperature gradient of the saturated adiabatic in the clouds. In order to explain this phenomenon somewhat further, we shall consider

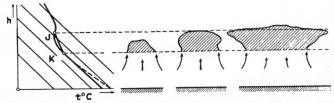

Fig. 28.—The clouds formed at height K enter a layer in which the temperature gradient is greater than that of the "damp adiabatic"—which is the law of expansion for the cloud. They therefore rise till they reach an inversion layer J.

in Fig. 28 an approximately adiabatic distribution. Let us suppose that on a hot summer day the surface of the earth is strongly heated; then, as always happens when the surface temperature rises, the layer above the surface becomes unstable (dotted curve) and gives rise to ascending and descending air currents. If the air has a relatively large humidity (as on a sultry day), the uprising currents soon reach the height of condensation (K).

If the dry air at this height is only slightly stable so that it is colder and therefore heavier than the uprising damp air, the clouds will increase in size through continually increasing condensation (according to the saturated adiabatic) up to a very stable air layer J (where the air is warmer and there is an inversion). On account of its inertia the cloud will overshoot sometimes and penetrate the region above J. If the supply of moist air is sufficient, the process described above can go on for a long time and on a large scale, causing a heat thunderstorm. If the condensation is sufficient, the water set free is precipitated as rain. Should the supply of damp air fail, the clouds detach themselves from the level at which they were formed and spread out under the inversion layer in the form of a cloud sheet. Cumulus clouds can thus change into stratus clouds.

Fig. 29.—A cold air mass breaks in over a long front. The warm air is pushed up and condensation appears. (Cold-front thunderstorm.)

We shall now consider the possibility of cloud formation in a large plain region without mountains. If the barometer is low over part of this region, there is a concentric inflow of air which lifts the air already there. Warm air comes in from the sides and the temperature gradient approximates to an adiabatic distribution. If condensation takes place at a certain height the equilibrium of the damp air can become unstable. As the air rises further, the condensation increases rapidly and produces long-continued rainfall. This process can go on for many days (rainy weather region).

A region of high pressure provides exactly opposite conditions. The air moves out radially, the height of the air layer decreases, and the temperature gradient becomes smaller, so that the distribution becomes stable. As the whole mass of air sinks downward the up currents produced by the surface heat are kept down so that the condensation has no important effect (fair weather region). We will not here enter into a consideration of the fundamental effect of the earth's rotation upon these meteorological processes.

In conclusion we shall discuss quite briefly a second kind of storm formation, which may be described as a storm caused by an upset of the weather. Here cold air breaks in under the warm air and pushes it up with consequent condensation (Fig. 29); this phenomenon nearly always takes place along an extensive front (cold-front thunderstorm). After the storm the temperature is much lower; this is not a consequence of the storm but of the inrush of cold air which caused the storm. The storm must be regarded as a consequence of the inrush, increased in violence by the resultant condensation.

CHAPTER III

STATIC LIFT ON GAS-FILLED AIRCRAFT

19. Pressure on the Balloon Wall.—Gas-filled aircraft can be divided into balloons and airships. Since the laws of static lift are the same for both, we shall limit the discussion to the behavior of the balloon.

A balloon consists essentially of a spherical gas-impermeable envelope, fitted with a "sleeve" for filling it with gas and with a basket containing apparatus and ballast (Fig. 30).

Since the gas expands with increasing height (about 0.3 per cent of the volume for every 100 ft) and the material of the envelope is unable to support the consequent forces, it is necessary that the filling sleeve should remain open during the journey in order to allow the expanding gas to escape. We shall now consider the equilibrium of a balloon completely filled with gas. The relation between pressure and height is given by Eq. (9a) on page 20.

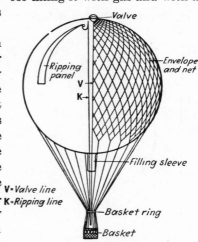

FIG. 30.—Diagrammatic representation of a balloon.

$$\frac{dp}{dz} = -\gamma,$$

where γ is the weight per unit volume.

We have to distinguish between two different weights per unit volume, that of the air γ_A round the balloon and that of the gas γ_G in the balloon. The difference in the value of γ_A immediately above and below the balloon is of the order of magnitude of the height of the balloon divided by the height of the atmosphere. With a balloon height of 60 ft and a uniform atmosphere of 5

miles, the difference is about 0.25 per cent and therefore **negligible**. We shall take a mean value for γ_A and for γ_G.

The pressure equation on integration gives

$$p_A = -\gamma_A z + \text{const.}$$

At $z = 0$, we have

$$p = p_0,$$

so that

$$p_A = p_0 - \gamma_A z.$$

Similarly for the gas in the balloon,

$$p_G = p_0 - \gamma_G z,$$

because at the opening of the sleeve $z = 0$, we have

$$p_G = p_A = p_0.$$

(In the case of airships there is an excess of internal pressure **so**

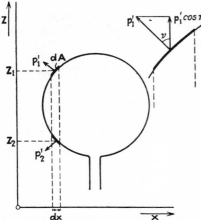

that an additive constant occurs in the pressure equation for the gas.) The pressure difference on the balloon envelope therefore is

$$p' = p_G - p_A = (\gamma_A - \gamma_G)z. \quad (1)$$

The resulting force on an element dA of the surface is by Fig. 31

$$p'dA$$

of which the vertical component is

$$p'dA \cos \nu,$$

Fig. 31.—Forces on the balloon wall.

where $dA \cos \nu$ is the vertical projection on the horizontal plane, *i.e.*, $dxdy$.

20. Lift of a Gas-filled Balloon.—The vertical component of the force on an element of surface dA has been found to be $p'dxdy$. The total resultant force, *i.e.*, the lift, is obtained by integration over the whole surface of the balloon:

$$L = \int\int (p_1' - p_2')dxdy,$$

or from (1)

$$L = (\gamma_A - \gamma_G)\int\int (z_1 - z_2)dxdy.$$

Therefore

$$L = (\gamma_A - \gamma_G) \cdot V = Q_A - Q_G, \tag{2}$$
$$\text{(Law of Archimedes)}$$

if we designate with $Q_G = \gamma_G V$ the weight of the gas and with $Q_A = \gamma_A V$ the weight of air displaced.

The equation $L = Q_A - Q_G$ also holds for non-homogeneous gas in the balloon. For if γ_G is a function of z,

$$p_G = p_0 - \int_0^z \gamma_G dz$$

and

$$p' = p_A z - \int_0^z \gamma_G dz.$$

Therefore

$$L = \iint p' dx dy = \gamma_A V - \iiint \gamma_G dx dy dz$$

and

$$L = Q_A - Q_G.$$

21. Effect of Temperature on Lift.—By introducing the temperature we can put the expression for the lift into another form. From Art. 10,

$$\gamma = \frac{1}{v},$$

which with the aid of the equation of state becomes

$$\gamma = \frac{p}{BT}. \tag{3}$$

Thus the lift L is given from Eq. (2)

$$L = pV\left(\frac{1}{B_A T_A} - \frac{1}{B_G T_G}\right).$$

The specific gravity of the gas (referred to air) can be used instead of the gas constant. If the temperature of the gas, T_G, is assumed equal to that of the air, T_A, then the specific gravity, σ, of the gas, referred to air at the same temperature, is

$$\sigma = \frac{\gamma_G}{\gamma_A} = \frac{\dfrac{p}{B_G T_G}}{\dfrac{p}{B_A T_A}} = \frac{B_A}{B_G}.$$

The gas constants are therefore inversely proportional to the weights per unit volume. If the gas and the air are at different temperatures,

$$\frac{\gamma_G}{\gamma_A} = \frac{B_A}{B_G}\frac{T_A}{T_G} = \sigma\frac{T_A}{T_G}. \tag{4}$$

If we substitute $B_G = B_A/\sigma$ in the above equation for the lift L we get

$$L = \frac{pV}{B_A}\left(\frac{1}{T_A} - \frac{\sigma}{T_G}\right)$$

or

$$L = \gamma_A V\left(1 - \sigma\frac{T_A}{T_G}\right)$$

or

$$L = Q_A\left(1 - \sigma\frac{T_A}{T_G}\right).$$

This is the fundamental formula for the lift on gas-filled aircraft.

22. Equilibrium of Forces on a Balloon.—Besides the lift force due to pressure differences there are a number of other forces to be taken into account.

1. The weight of the balloon
2. The weight of the crew $\Big\}$ constant weight Q_1,

3. The weight of the ballast (mostly in the form of $\Big\}$ variable
   sacks of sand of about 30 lb each) $\Big\}$ weight Q_2.

If the lift is in equilibrium with these forces,

$$L = Q_1 + Q_2.$$

If the lift is altered (by the sun shining on the balloon), or if the amount of ballast is changed (by throwing out sand), a resultant force R acts on the balloon, part of which produces an acceleration and part of which overcomes the resistance of the air:

$$R = L - (Q_1 + Q_2). \tag{5}$$

Airships, in addition to the static lift, are also acted upon by dynamic forces caused by the inclined position of the airship due to rudder adjustments.

We shall now investigate the alteration of force or of height caused by a change in the various quantities concerned; these changes can be produced by:

1. Loss of gas,
2. Loss of ballast,
3. Alteration of air temperature,
4. Alteration of gas temperature.

It is of importance for the following investigation to distinguish between two different states of the balloon:

1. Taut state,
2. Limp state.

In the taut state the balloon is completely filled with gas, so that the *volume* of gas in the balloon remains constant with all changes of position.

In the limp state only a part of the balloon is filled with gas, so that either the envelope lies in folds or else air enters through the filling sleeve and occupies the lower part of the envelope; in this state the *weight* of the gas remains constant.

In the taut state a balloon can rise only by losing gas, but in the limp state it can rise without loss of gas. The various heights to which a balloon in the taut state can rise (H_1, H_2, H_3, Fig. 32) are called "taut heights." By plotting the height as a

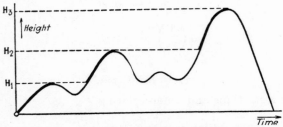

FIG. 32.—Change of height of a balloon. The thick lines correspond to the full or taut state, and the other lines to the limp state.

function of the time, a curve of the kind shown in Fig. 32 will be obtained. The thick parts of the curve correspond to the taut states while for the rest of the time the balloon was in the limp state.

23. Stability of a Balloon in Taut State under Adiabatic Conditions.—We shall first investigate the stability under adiabatic conditions (no change of temperature by radiation) and afterward determine the influence exerted by changes of temperature, caused by emission or absorption of heat. In both cases we consider first the taut, then the limp state.

Equation (5) in combination with (2) gives

$$R = V(\gamma_A - \gamma_G) - Q, \tag{6}$$

where Q is the total weight ($Q = Q_1 + Q_2$).

To calculate the change in the resultant force with height we find dR/dh at constant volume (since we are considering the balloon as taut); we assume at first that the total weight Q is constant. Then

$$\frac{dR}{dh} = V\frac{d}{dh}(\gamma_A - \gamma_G)$$

$$\frac{dR}{dh} = V\frac{dp}{dh}\cdot\frac{d}{dp}(\gamma_A - \gamma_G),$$

or, since

$$\frac{dp}{dh} = -\gamma_A,$$

we have

$$-\frac{dR}{dh} = V\gamma_A\frac{d}{dp}(\gamma_A - \gamma_G). \tag{7}$$

If we now assume that the air distribution is a polytrope of the form

$$pv^{n_A} = \text{const.}$$

or

$$p\left(\frac{1}{\gamma_A}\right)^{n_A} = \text{const.}$$

or

$$\gamma_A = \text{const. } p^{\frac{1}{n_A}},$$

or taking logarithms and differentiating,

$$d(\log \gamma_A) = \frac{d\gamma_A}{\gamma_A} = \frac{1}{n_A}\frac{dp}{p},$$

then

$$\frac{d\gamma_A}{dp} = \frac{\gamma_A}{n_A p}.$$

The expression $d\gamma_G/dp$ can be derived similarly assuming the adiabatic relation

$$pv^{k_G} = \text{const.},$$

where k_G is about 1.35 for coal gas and 1.40 for hydrogen. We explicitly have excluded any heat emission or absorption from the argument, so that k is taken instead of n. We get, by analogy with the previous analysis,

$$\frac{d\gamma_G}{dp} = \frac{\gamma_G}{k_G p}.$$

Substituting the expressions found for $d\gamma_A/dp$ and $d\gamma_G/dp$ in (7), we obtain

$$-\frac{dR}{dh} = \frac{V}{p}\gamma_A\left(\frac{\gamma_A}{n_A} - \frac{\gamma_G}{k_G}\right).$$

For rough estimates it is permissible to assume

$$n_A = k_G = n.$$

The formula then becomes very simple:

$$-\frac{dR}{dh} = \frac{V\gamma_A}{pn}(\gamma_A - \gamma_G)$$

or introducing $p/\gamma_A = h_0$ (height of the uniform atmosphere):

$$-\frac{dR}{dh} = \frac{V(\gamma_A - \gamma_G)}{h_0 n} = \frac{Q}{h_0 n}. \tag{8}$$

As an example of the practical application of this expression we shall find out how high a balloon will ascend in the taut state if a given quantity of ballast is thrown out. For equilibrium we have $R = 0$. If a quantity of ballast of weight ΔQ is thrown out, then, at that instant, $\Delta R = -\Delta Q$. By expanding h as a function of R in a Taylor's series and using only the first term, we find

$$\Delta h = \frac{dh}{dR}\Delta R,$$

or, since at the moment of releasing the ballast $\Delta R = -\Delta Q$, and since from (8) we can write

$$\frac{dh}{dR} = -\frac{nh_0}{Q},$$

we obtain

$$\Delta h = \frac{h_0 n}{Q}\Delta Q. \tag{9}$$

(The ballast formula)

If $Q = 2,000$ lb and $\Delta Q = 20$ lb, *i.e.*, 1 per cent of the total weight, and assuming $t = 0°C$, $n = 1.40$ (balloon filled with hydrogen) and $h_0 = 5$ miles (see page 30),

$$\Delta h = \frac{5 \times 5,280 \times 1.4 \times 20}{2,000} = 370 \text{ ft.}$$

It is seen that a considerable loss of ballast produces only a small change in height, which shows that the taut state is very stable.

24. Stability of a Balloon in the Limp State under Adiabatic Conditions.—Since in the limp state the available volume is only partly filled with gas, the balloon can move upward without gas escaping from the filling sleeve. The total weight of gas Q_G remains constant as long as the balloon is limp.

Inserting $V = Q_G/\gamma_G$ into (6), we get

$$R = \frac{Q_G}{\gamma_G}(\gamma_A - \gamma_G) - Q = Q_G\left(\frac{\gamma_A}{\gamma_G} - 1\right) - Q.$$

If we now find dR/dh with Q_G constant, we obtain

$$\frac{dR}{dh} = Q_G \frac{dp}{dh} \times \frac{d}{dp}\left(\frac{\gamma_A}{\gamma_G} - 1\right).$$

But

$$\frac{dp}{dh} = -\gamma_A,$$

so that

$$\frac{dR}{dh} = -\frac{Q_G\gamma_A}{\gamma_G{}^2}\left(\gamma_G \frac{d\gamma_A}{dp} - \gamma_A \frac{d\gamma_G}{dp}\right).$$

As in Art. 23 this equation can be changed to

$$\frac{dR}{dh} = -\frac{Q_G\gamma_A{}^2}{\gamma_G p}\left(\frac{1}{n_A} - \frac{1}{k_G}\right).$$

Substituting $p/\gamma_A = h_0$,

$$\frac{dR}{dh} = -\frac{V\gamma_A}{h_0}\left(\frac{1}{n_A} - \frac{1}{k_G}\right).$$

Since in the equilibrium state

$$Q = V(\gamma_A - \gamma_G),$$

we obtain

$$-\frac{dR}{dh} = \frac{Q\gamma_A}{h_0(\gamma_A - \gamma_G)}\left(\frac{1}{n_A} - \frac{1}{k_G}\right).$$

Equation (4) states

$$\frac{\gamma_A}{\gamma_A - \gamma_G} = \frac{1}{1 - \dfrac{\gamma_G}{\gamma_A}} = \frac{1}{1 - \sigma\dfrac{T_A}{T_G}}.$$

Substituting this in our expression for dR/dh, we find

$$-\frac{dR}{dh} = \frac{Q}{h_0\left(1 - \sigma\dfrac{T_A}{T_G}\right)}\left(\frac{1}{n_A} - \frac{1}{k_G}\right). \tag{10}$$

When $n_A = k_G$, dR/dh is zero, *i.e.*, a change in the height h produces no variation of R, so that there is a state of neutral equilibrium. For $n_A = 1.40$ and $k_G = 1.35$, or whenever $n_A > k_G$, $dR/dh > 0$, which means that the resultant force increases with the height and the equilibrium is unstable. Equilibrium is stable only when $n_A < k_G$.

As yet we have investigated the stability of a balloon only under adiabatic conditions. We shall now consider the effect of temperature changes caused by transfer of heat to or from the

balloon, *i.e.*, temperature changes at constant pressure such as are set up in particular by absorption or emission of radiation.

25. Effect of Temperature Changes at Constant Pressure on a Balloon in the Taut State.—Starting from Eq. (6) and using (3) and (4) we can write

$$R = \frac{Vp}{B_A}\left(\frac{1}{T_A} - \frac{\sigma}{T_G}\right) - Q.$$

If we now differentiate R partially with respect to the temperature T of the gas, keeping V and p constant according to our assumptions of taut state and constant pressure, we get

$$\frac{\partial R}{\partial T_G} = \frac{Vp}{B_A}\frac{\sigma}{T_G{}^2},$$

and using (3) and (4)

$$\frac{\partial R}{\partial T_G} = \frac{V\gamma_G}{T_G}.$$

Since

$$V\gamma_G = \frac{V(\gamma_A - \gamma_G)}{\gamma_A - \gamma_G}\gamma_G = \frac{Q}{\frac{\gamma_A}{\gamma_G} - 1} = \frac{Q\frac{\gamma_G}{\gamma_A}}{1 - \frac{\gamma_G}{\gamma_A}} = \frac{Q\sigma\frac{T_A}{T_G}}{1 - \sigma\frac{T_A}{T_G}}, \qquad (11)$$

we can also write

$$\frac{\partial R}{\partial T_G} = \frac{Q}{T_G}\frac{\sigma\frac{T_A}{T_G}}{1 - \sigma\frac{T_A}{T_G}}.$$

Similarly differentiating with respect to T_A, we find

$$\frac{\partial R}{\partial T_A} = -\frac{Vp}{B_A T_A{}^2} = -\frac{V\gamma_A}{T_A} = -\frac{Q}{T_A\left(1 - \sigma\frac{T_A}{T_G}\right)}.$$

For finite small differences of temperature of the gas (ΔT_G) and of the air (ΔT_A), we get

$$\frac{\Delta R}{Q} = \frac{\sigma\frac{\Delta T_G}{T_G}\cdot\frac{T_A}{T_G} - \frac{\Delta T_A}{T_A}}{1 - \sigma\frac{T_A}{T_G}}, \qquad (12)$$

because

$$\Delta R = \frac{\partial R}{\partial T_G}\Delta T_G + \frac{\partial R}{\partial T_A}\Delta T_A.$$

A simple approximate formula can again be obtained by assuming the gas and air to be of the same temperature,

$$T_A = T_G = T.$$

Then

$$\frac{\Delta R}{Q} = \frac{\sigma \Delta T_G - \Delta T_A}{T(1 - \sigma)}$$

and, using the ballast formula (9),

$$\Delta h = n h_0 \left(\frac{\sigma}{1 - \sigma} \frac{\Delta T_G}{T} - \frac{1}{1 - \sigma} \cdot \frac{\Delta T_A}{T} \right). \tag{12a}$$

This equation gives the relation between the changes of height in the taut state and temperature at constant volume. The first term is the most important, since $\sigma/1 - \sigma$ alters rapidly with σ; for example, with coal gas

$$\frac{\sigma}{1 - \sigma} = \frac{0.42}{0.58} = 0.73,$$

while for hydrogen

$$\frac{\sigma}{1 - \sigma} = \frac{0.07}{0.93} = 0.075,$$

which is almost one-tenth of the value for coal gas.

As an example let us calculate the change in height for a temperature rise of 1°C, taking (1) $T = 290°(t = 17°C)$, (2) $h_0 = 5$ miles, (3) $n_A = 1.35$ (coal gas). Substituting these values in our simplified formula (12a) we find $\Delta h = 87$ ft. The corresponding conditions for hydrogen make $\Delta h = 9$ ft. Thus a taut balloon filled with hydrogen is much more stable as regards height alterations due to temperature changes than one filled with coal gas. The great difference between the two can be understood from the difference in the specific gravities; on expansion the loss of coal gas is equivalent to a much greater loss of ballast than that caused by the expansion of hydrogen.

26. Effect of Temperature Changes at Constant Pressure on a Balloon in the Limp State.—Substituting $V = Q_G/\gamma_G$ into (6), we obtain

$$R = Q_G \left(\frac{\gamma_A}{\gamma_G} - 1 \right) - Q,$$

or from (4)

$$R = Q_G \left(\frac{T_G}{\sigma T_A} - 1 \right) - Q.$$

To find the dependence of the resultant force on the temperature change, we again differentiate R partially with respect to T_G and T_A, the weight Q_G of the gas being constant since the balloon is limp.

$$\frac{\partial R}{\partial T_G} = \frac{Q_G}{\sigma T_A},$$

or from (11)

$$\frac{\partial R}{\partial T_G} = \frac{Q}{\left(1 - \sigma \dfrac{T_A}{T_G}\right) T_G},$$

and, as in Art. 25, we get from (4),

$$\frac{\partial R}{\partial T_A} = -\frac{Q_G T_G}{\sigma T_A{}^2} = -V \gamma_G \frac{\gamma_A}{\gamma_G T_A} = -\frac{Q_A}{T_A} = -\frac{Q}{\left(1 - \sigma \dfrac{T_A}{T_G}\right) T_A}.$$

For finite temperature differences of the balloon gas and of the air we then get in an analogous manner as in Art. 25:

$$\frac{\Delta R}{Q} = \frac{\dfrac{\Delta T_G}{T_G} - \dfrac{\Delta T_A}{T_A}}{1 - \sigma \dfrac{T_A}{T_G}}. \tag{13}$$

In distinction to Eq. (12), coal gas and hydrogen do not behave very differently since (13) does not contain $\sigma/1 - \sigma$. The change in the gas temperature acts only through the resulting change in V.

Combining (10) with (13):

$$\Delta h = \frac{\dfrac{\Delta T_G}{T_G} - \dfrac{\Delta T_A}{T_A}}{\dfrac{1}{n_A} - \dfrac{1}{k_G}} h_0. \tag{14}$$

This gives for the limp state the relation between the changes of temperature and height; it is to be noted that the change in height is independent of σ.

As an example we shall suppose that the temperature of the coal gas rises by 1°C while the air temperature ($T_A = 290°$) remains constant,

$$\Delta T_G = 1°C (k_G = 1.35),$$
$$\Delta T_A = 0,$$
$$T_G = T_A = 290°,$$

corresponding to $h_0 = 5$ miles.

Substituting in (14), different values for n_A (for different air distributions) we obtain the following table:

$k_G = 1.35$	$\dfrac{1}{n_A} - \dfrac{1}{k_G}$	Δh, ft
$n_A = 1.35$	0	∞
$n_A = 1.20$	0.092	1,050
$n_A = 1.00$	0.259	370

27. Causes of Heat Changes; Behavior of Balloon during Travel.—Changes of temperature at constant pressure may take place during the daytime by the sun shining on the balloon, and during the night by radiation from the balloon, and also by the cooling produced by vertical motion (about 3 to 15 ft/sec, horizontal motion relative to the air is practically non-existent). The heat emission produced by vertical motion is proportional to the velocity, while the resistance is proportional to the square of the velocity.

Since in the daytime the balloon is above the temperature of its surroundings, every upward motion is diminished by cooling due to its motion, while a downward motion is accelerated. In the limp state, therefore, it rises slowly and uniformly as long as the sun shines but falls quickly as soon as clouds appear. At night the converse is true; the balloon is then colder than the surrounding air and it rises quickly and sinks slowly. It may be mentioned that for long sustained journeys it is correct to remain in the taut state during the day and to seek a stability layer (ground inversion of Art. 18) during the night.

As we have mentioned, the resistance or drag to vertical motion D is proportional to the second power of the velocity. Applying Newton's second law to the vertical motion, we obtain, neglecting the effect of cooling by the motion,

$$m\frac{d^2h}{dt^2} = R(h) \mp D,$$

where

$$D = \text{const.}\ \frac{\gamma_A}{g}A\left(\frac{dh}{dt}\right)^2.$$

The negative sign is for ascent and the positive for descent.

This differential equation of the second order and second degree can be transformed to one of the first order if R is inde-

pendent of h by introducing the variable dh/dt. The problem is more complicated if cooling due to movement takes place, since R is then not only dependent on h but is also given by a differential equation. The mass of air carried along by the

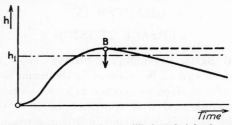

FIG. 33.—The balloon rises above the equilibrium height h_1 on account of its inertia. It then falls to earth slowly if no ballast is thrown out.

balloon must also be taken into account; its result is to increase the effective mass of the balloon so that $m = 1.5Q/g$ approximately.

After loss of ballast the balloon overshoots its equilibrium position on account of inertia and would sink to earth again because of the resulting downward force if the air were not sufficiently stable Fig. (33). It is therefore necessary at the end of the period of upward motion (B) to throw out a little ballast (about 1 lb) in order to bring the balloon into equilibrium.

Before landing, brake ballast must be thrown out. A part of the brake ballast is formed

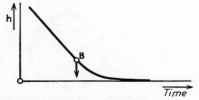

FIG. 34.—The effect of throwing out ballast shortly before landing.

by the drag rope. Another purpose of the drag rope is to pull the balloon round so that the ripping panel, which is provided for the final opening of the balloon, is placed behind and upward. The purpose of throwing out ballast shortly before a landing (point B, Fig. 34) is to diminish the shock of landing. Shortly before contact, the ripping cord is pulled and the gas escapes rapidly from the big opening. The landing is then accomplished.

CHAPTER IV

SURFACE TENSION

28. Physical Basis.—Free liquid surfaces show a variety of phenomena which can all be reduced to the same cause, namely, the tendency of a surface to become as small as possible. This fact, which is connected with the presence of a tension in the surface skin similar to the tension in a thin stressed membrane, can be demonstrated very strikingly by the following simple experiment (Fig. 35). If a wire ring, to which a thin loop of thread is attached, is dipped in a soap solution (to which a few drops of glycerine have been added to prevent rapid evaporation) it will be covered by a very thin film. The loop of thread will

FIG. 35.—The effect of surface tension in a thin fluid skin.

take up no particular position in this film. If the part of the film inside the thread be pierced, the loop will be pulled out into a circle by the surface tension. The thin fluid film has thus attained its smallest area.

The fact that a liquid under no external forces will take up a spherical shape can also be ascribed to the property of the surface to have a minimum area for a given volume. This particular behavior of a liquid surface is connected with the unsymmetrical molecular force on the molecules in the surface. While inside the liquid the molecular forces compensate each other, the molecules on the surface experience a force directed toward the inside so that the surface molecules are prevented from escaping. In this way the surface shows a tendency to become as small as possible.

Surface phenomena occur not only at free surfaces but also at interfaces between fluids, for example, between water and oil. The tensions at the surfaces of contact between two media lead to phenomena similar to the surface tensions at the interface between a liquid and gas.

In certain cases, however (when the liquids are miscible like water and alcohol), there is a force of compression instead of tension. Such surface skins are however in unstable equilibrium and generally not observable; only by taking special precautions, such as pouring the alcohol on to the water, is it possible momentarily to see the surface skin which has a crumpled appearance.

29. Relation between Surface Tension and Pressure Difference across a Surface.—If we consider any part of a liquid surface, the surface tension is the same at all parts of the surface skin, and at every point its direction is in the plane tangential to the surface. We define the surface tension as the tangential force per unit length of a cut in the liquid surface. The surface tension C of water is 74 dynes/cm or 0.00042 lb/in.

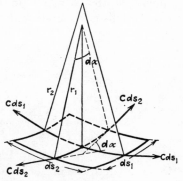

Fig. 36.—Surface tension on a curved surface.

To derive the relation between surface tension and the resultant force perpendicular to the surface we shall consider a rectangular element of the surface (Fig. 36) and take r_1 and r_2 as the radii of curvature for two directions at right angles to each other in the surface. If C is the surface tension, and if ds_1 and ds_2 are the lengths of the sides, the forces Cds_1 and Cds_2 will act on the sides of the element. If we draw the diagram of the forces perpendicular to ds_2 then, since $ds_1 = r_1 d\alpha$, the resultant perpendicular force is given by

$$Cds_2 d\alpha = Cds_2 \frac{ds_1}{r_1}.$$

Similarly the resultant force perpendicular to ds_1 is

$$Cds_1 \frac{ds_2}{r_2}.$$

For equilibrium, the sum of these two forces must clearly be balanced by a pressure difference; the force, due to this pressure difference on the surface of area $ds_1 ds_2$ is

$$\Delta P = \Delta p\, ds_1 ds_2.$$

Thus the pressure difference per unit area is

$$\Delta p = C\left(\frac{1}{r_1} + \frac{1}{r_2}\right). \tag{1}$$

The pressure difference between the two sides of a liquid surface is proportional to the sum of the reciprocals of the radii of curvature in two mutually perpendicular directions; the pressure is always greater on the concave side of the surface. It is to be noted that the expression $(1/r_1 + 1/r_2)$ is invariant with respect to rotation of the coordinates, so that Eq. (1) is independent of the direction in which the rectangular element is taken. This is in agreement with the physical point of view.

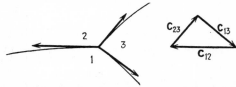

Fig. 37.—Surface tensions at the place of contact of three fluids. The angles are fully determined by the sides of the triangle.

30. Surface Tension at Place of Contact between Several Media.

—Up to the present we have considered only the behavior of the free surface of a fluid, *i.e.*, of a fluid bounded by a gas. The phenomena have been explained by the supposition that the molecules of the liquid are influenced by the cohesion forces of the liquid alone, since the corresponding gaseous forces on these liquid molecules are negligible. The resultant force directed inward was seen to be the cause of the surface phenomena.

Conditions are different however if the liquid is bounded by other liquids or by solids; the surface of the liquid will then be acted on by forces from the contiguous media and the surface tension will be reduced. It is to be noted that the surface tension $C_{1,2}$ between two bodies 1 and 2 is not even approximately equal to the difference between the surface tensions of the bodies and air $(C_1 - C_2)$.

When three bodies (of which one may be air) meet, the surface tensions are related as shown in Fig. 37. When two fluids

1 and 2 (of which again one may be air) come into contact on the surface of a solid body 3, the two fluids meet the solid surface at an angle α at the point of contact P (Fig. 38). Since the equilibrium in the vertical direction is given by the presence of the solid, it is only necessary that the resultant of the horizontal components of the surface tensions be zero.

$$C_{1,2} \cos \alpha = C_{2,3} - C_{1,3}.$$

Knowing the surface tensions it is therefore possible to calculate the angle of contact, or by measuring this angle to calculate one of the three surface tensions.

If $C_{2,3} - C_{1,3} < 0$, then $\alpha > \pi/2$ and we say that the liquid 1 does not wet the solid (mercury on glass is an example; see

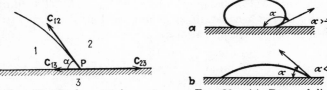

FIG. 38.—Surface tensions at the place of contact of two fluids (one of which may be air) and a rigid body.

FIG. 39.—(a) Drop of liquid "not wetting" the surface, $\alpha > \pi/2$. (b) Drop of liquid "wetting" the surface, $\alpha < \pi/2$.

Fig. 39a); if $C_{2,3} - C_{1,3} < C_{1,2}$, the above equation cannot be satisfied at any angle, there can be no equilibrium, and the point P moves to the right so that the liquid creeps along continually. This phenomenon is shown by certain oils on glass or metal, or by oil drops which when allowed to fall on water finally cover the whole surface with a thin layer.

31. Surface Effects under the Action of Gravity.—We have not yet considered the effect of external forces, especially gravitational forces on the phenomena of the surface action of liquids, and we shall now investigate the shape taken up by the surface of separation between two media (one of which may be air) under the action of gravity. We shall use a rectangular coordinate system with the positive direction of z upward.

Let the pressure at any point in the liquid 1, (weight per unit volume γ_1) be p_1, and the pressure in the liquid 2 (of weight per unit volume γ_2) be p_2, then from page 20

$$p_1 = p_0 - \gamma_1 z$$

and

$$p_2 = p_0 - \gamma_2 z,$$

where p_0 is an integration constant which we shall assume to be the same for both liquids. This means that when $z = 0$, the pressure difference $p_2 - p_1 = 0$, and the origin of the coordinate system therefore lies in the plane at which $p_1 = p_2$.

The pressure difference $p_2 - p_1 = \Delta p$ at the surface of separation of the two liquids is

$$\Delta p = (\gamma_1 - \gamma_2)z.$$

Since this pressure difference is connected with the radius of curvature by Eq. (1), we have

$$\frac{1}{r_1} + \frac{1}{r_2} = \frac{\gamma_1 - \gamma_2}{C_{1,2}}z. \tag{2}$$

The radii are positive if the interface is concave in the upward direction, the liquid 2 being on top of the liquid 1. The quotient $(\gamma_1 - \gamma_2)/C_{1,2}$ is a constant of geometrical significance with the dimensions of the reciprocal of an area:

$$\frac{\text{Force}}{\text{Volume}} \times \frac{\text{length}}{\text{force}} = \frac{1}{\text{length}^2}.$$

Fig. 40.—Analogy between a thin bent elastic wire and the shape of the free surface under the action of surface tension.

If the radii of curvature r_1 and r_2 are expressed in the above Eq. (2) in terms of x and y, a differential equation will be obtained in which z is a function of x and y. The solution of this equation gives the surface of separation, the integration constants being fixed by the boundary conditions. The problem is considerably simplified if $1/r_2$ is zero, as is the case when two fluids are in contact along a flat wall. The solution of the differential equation then leads to the same elliptic functions as occur in the problem of the bending of thin wires (examples are shown in Fig. 40). With surfaces having rotational symmetry, as for example cylinders, the solution of the differential equation cannot be given explicitly.

32. Capillarity.—If we dip a narrow glass tube into water, we know from the above—since water wets glass very well—that the water will make a very sharp angle of contact, so that the surface

will be very curved provided the tube is sufficiently narrow. The result of this, however, is to produce, as we have seen in Art. 29, a force ΔP upwards; the water therefore rises in the tube until the weight of the column of water equals ΔP (Fig. 41).

If the inside of the tube has been wetted so that the angle of contact α is zero, then, from (2), if a is the radius of the tube and the meniscus is assumed as an approximation to have a hemispherical shape,

$$\frac{2}{a} = \frac{\gamma_1 - \gamma_2}{C_{1,2}}z,$$

and, therefore,

$$z = \frac{2C_{1,2}}{(\gamma_1 - \gamma_2)a}.$$

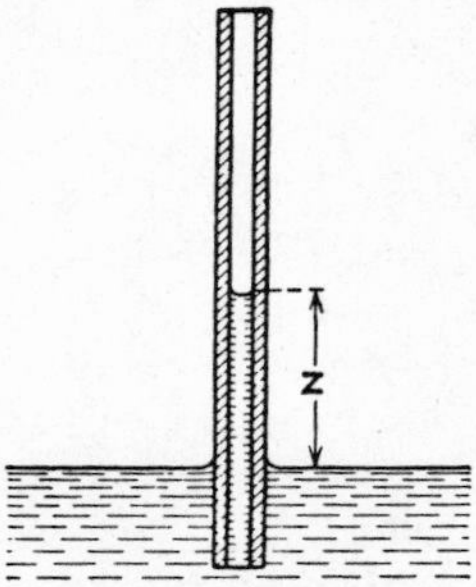

Fig. 41.—Capillary tube.

This relation can also be derived directly if the surface force around the perimeter $2\pi a$ of the tube is regarded as balancing the weight $\pi a^2 z(\gamma_1 - \gamma_2)$ of the column of liquid:

$$2\pi a C_{1,2} = \pi a^2 z(\gamma_1 - \gamma_2).$$

Therefore

$$z = \frac{2C_{1,2}}{(\gamma_1 - \gamma_2)a}.$$

This derivation also suggests a refinement in our formula; z should be taken as the height at which the volumes are equal as shown in Fig. 42.

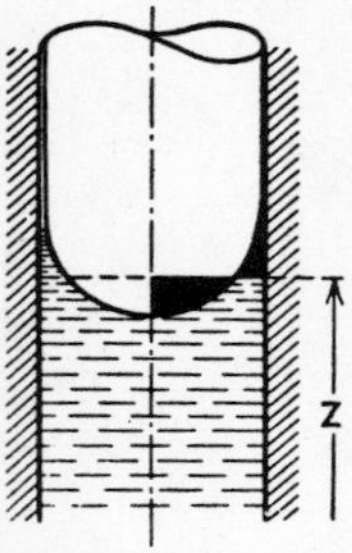

If the capillary tube is not wetted, then, if α is the angle between the liquid and glass,

$$2\pi a C_{1,2} \cos \alpha = \pi a^2 z(\gamma_1 - \gamma_2)$$

or

$$z = \frac{2C_{1,2} \cos \alpha}{(\gamma_1 - \gamma_2)a}.$$

Fig. 42.—Meniscus in a capillary tube.

As an example let $a = 0.020$ in., then, taking $C_{1,2} = 0.00042$ lb./in., the height to which the water will rise is

$$z = \frac{2 \times 0.00042}{0.036 \times 0.020} = 1.17 \text{ in.}$$

PART II

KINEMATICS OF LIQUIDS AND GASES

CHAPTER V

METHODS OF DESCRIPTION

33. Lagrangian Method.—This part is concerned with the relations between the position of the various fluid particles and the time, whereby the fluid is treated as a continuum. Since the dimensions of the fluid particles to be considered are always large compared with the mean free path of the molecules, such a treatment is quite valid as has been shown in Art. 1.

Two methods of describing fluid motion have been devised. One is concerned with what happens to the individual fluid particles in the course of time, what paths they describe, what velocities or accelerations they possess, and so on. The other method attempts to find out what happens at a given point in the space filled by the fluid and what the velocities, etc., are at the different points in this space.

We shall first describe the first method and in order to distinguish between the separate fluid particles we must so to speak label them or provide them with names. We do this by setting up some system of coordinates at the time $t = t_0$ (a real or fictitious starting point). In this way to every particle in the continuum are assigned three parameters (a, b, c). These three letters are the name of the fluid particle concerned, because it retains them throughout the whole investigation. The fluid, every part of which is designated by its appropriate three letters, will now be referred to a rectangular coordinate system. The motion of the fluid is completely specified if the vector $\mathbf{r}$ (representing the position of a particle) or the three corresponding coordinates x, y, z are given as a function of the time. This is expressed by the equation

$$\mathbf{r} = \mathbf{F}(a, b, c, t,) \quad \text{or} \quad \begin{cases} x = F_1(a, b, c, t) \\ y = F_2(a, b, c, t) \\ z = F_3(a, b, c, t) \end{cases} \tag{1}$$

Without limiting the generality of this method, the three letters (a, b, c) representing the name of the fluid particle can also be

considered as Cartesian coordinates. Every fluid particle is then uniquely determined by a vector

$$\mathbf{s} = \mathbf{i}a + \mathbf{j}b + \mathbf{k}c,$$

where $\mathbf{i}$, $\mathbf{j}$, $\mathbf{k}$ are the unit vectors along the x-, y-, z-axes. In this interpretation the x, y, z system, or as we shall call it the **r**-space, is fixed at some time $t = t_0$ and every fluid particle is denoted by the vector of position $\mathbf{s}$ which it has at time $t = t_0$. This fictitious space, which serves only to name the various particles, we shall call the **s**-space; in general it coincides with the **r**-space only at time $t = t_0$.

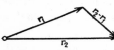

FIG. 43.—Vectorial position diagram.

The above equation, (1), therefore obtains the form

$$\mathbf{r} = \mathbf{F}(\mathbf{s}, t),$$

where $\mathbf{s}$ (the name of the particle) is constant for the particle concerned.

In this method of description therefore the velocity of a fluid particle $\mathbf{s}$ (Fig. 43) is given by

$$\mathbf{w} = \lim_{t_2 - t_1 = 0} \frac{\mathbf{r}_2 - \mathbf{r}_1}{t_2 - t_1} = \left(\frac{\partial \mathbf{r}}{\partial t}\right)_{\mathbf{s}} = \dot{\mathbf{r}}_{\mathbf{s}};$$

or

$$u = \left(\frac{\partial x}{\partial t}\right)_{\mathbf{s}} = \dot{x}_{\mathbf{s}},$$

$$v = \left(\frac{\partial y}{\partial t}\right)_{\mathbf{s}} = \dot{y}_{\mathbf{s}},$$

$$w = \left(\frac{\partial z}{\partial t}\right)_{\mathbf{s}} = \dot{z}_{\mathbf{s}}.$$

The subscript $\mathbf{s}$ means that the differentiation must be carried out with $\mathbf{s}$ constant.

Correspondingly for the acceleration,

$$\mathbf{a} = \left(\frac{\partial^2 \mathbf{r}}{\partial t^2}\right)_{\mathbf{s}} = \ddot{\mathbf{r}}_{\mathbf{s}};$$

or

$$a_x = \left(\frac{\partial^2 x}{\partial t^2}\right)_{\mathbf{s}} = \ddot{x}_{\mathbf{s}},$$

$$a_y = \left(\frac{\partial^2 y}{\partial t^2}\right)_{\mathbf{s}} = \ddot{y}_{\mathbf{s}},$$

$$a_z = \left(\frac{\partial^2 z}{\partial t^2}\right)_{\mathbf{s}} = \ddot{z}_{\mathbf{s}}.$$

This apparently very convenient method proves to be very troublesome and difficult in practice, when solutions to definite problems have to be found, though it is very powerful, to be sure, in the few cases where it can be carried through. This method is called the "Lagrangian" method because Lagrange more than anyone else worked out its calculation; the method was previously known, however, to Euler. Usually the elaborate result given by this method is more than one requires to know. In most cases, at least in homogeneous fluids, the behavior of individual fluid particles is not of particular interest; one only wants to know the state of motion and its alteration with the time at every point.

We shall now describe the second method which, while giving results which are less far reaching, lends itself much more easily to numerical investigation.

34. Eulerian Method and Its Connection with That of Lagrange.—The Eulerian method, named after the founder of hydrodynamics, answers the question: what is happening at a given time t at the various points $\mathbf{r}$ in a space filled by a fluid? The velocity at any point in the space can be written

$$\mathbf{w} = \mathbf{f}(\mathbf{r}, t);$$

or

$$\begin{aligned} u &= f_1(x, y, z, t) \\ v &= f_2(x, y, z, t). \\ w &= f_3(x, y, z, t) \end{aligned} \tag{2}$$

In order to find out subsequently what is happening to the individual fluid particles according to the Lagrangian method, we have a set of equations for each particle:

$$\left. \begin{aligned} \frac{dx}{dt} &= u \\ \frac{dy}{dt} &= v \\ \frac{dz}{dt} &= w \end{aligned} \right\}, \tag{3}$$

which, when combined with Eqs. (2), gives a set of three simultaneous differential equations of the first order showing the dependence of x, y, z on t. The solution of this set of equations involves three constants of integration which are necessary to fulfill the requirement that at a given time $t = t_1$ the coordinates

of the particle become $x = x_1,\ y = y_1,\ z = z_1$. These three integration constants, or the dependent quantities $x_1,\ y_1,\ z_1$, can be considered as initial coordinates $a,\ b,\ c$ of the fluid particle, so that we return to the equations of Lagrange:

$$x = F_1(a,\ b,\ c,\ t),$$
$$y = F_2(a,\ b,\ c,\ t),$$
$$z = F_3(a,\ b,\ c,\ t).$$

Therefore in principle the Lagrangian method of description can always be derived by (3) from the Eulerian method; but the difficulties of solving the three differential equations are generally so great that only a few applications of this method are known.

35. Streamlines and Paths of Particles; Steady Flow.—We shall now consider certain lines which are particularly suitable for describing the fluid motion.

Using the Eulerian method of description we have a velocity field corresponding to the equation $\mathbf{w} = f(\mathbf{r},\ t)$; this means that the velocity is given in direction and magnitude at every point in the space considered, and in general this velocity field alters with time. If we fix our attention to the time $t = t_1$, and plot curves such that their direction at each point agrees with the direction of the velocity at that point, then these curves map out the velocity direction at every point in the space. They are called "streamlines."

The method of Lagrange gives information about the paths of the separate fluid particles as a function of the time (path lines); the method of Euler gives, so to speak, a series of instantaneous photographs (or streamline diagrams) of the state of motion, but there is no relation between these streamlines and the individual fluid particles since in general the streamlines are formed by different fluid particles at different times.

The equation of the streamlines at a given time $t = t_1$ is (by definition):

Written vectorially

$$d\mathbf{r}\,\|\,\mathbf{w}$$

or

$$d\mathbf{r} \times \mathbf{w} = 0;[1]$$

In coordinates

$$dx : dy : dz = u : v : w$$

$$\left.\begin{array}{c} \\ \\ \\ \\ \end{array}\right\} \qquad (4)$$

---

[1] $\mathbf{a} \times \mathbf{b}$ is the vector product of the two vectors $\mathbf{a}$ and $\mathbf{b}$.

This equation does not differ from the set of Eqs. (3) at the time $t = t_1$, *i.e.*, the path lines are tangent to the streamlines at the time $t = t_1$ at the places where the particles happen to be situated just then.

A particularly important case is that in which t in (2) does not enter explicitly into the expression $\mathbf{w} = \mathbf{f}(\mathbf{r})$; this obviously means that the velocities at the different points in space are independent of time. Such motions are called "steady motions." In this case, (3) is identical with (4) (t appears as a parameter) so that we can say:

The paths of particles are identical with the streamlines in a steady flow.

36. Streak Lines.—In addition to these two important lines we shall consider yet a third kind, which is of particular experimental interest. Suppose that we want to know which fluid particles, *i.e.*, which $\mathbf{s}$ pass a given point $\mathbf{r}_1$ of the $\mathbf{r}$-space in the course of time. The fluid particles which satisfy this condition must, by the Lagrangian method, obey the equation

$$\mathbf{F}(\mathbf{s}, t) = \mathbf{r}_1 = \text{const.}$$

Or solving for $\mathbf{s}$,

$$\mathbf{s} = \mathbf{G}(\mathbf{r}_1, t).$$

For each instant t there will be an $\mathbf{s}$ which assumes the position of $\mathbf{r}_1$ at t, *i.e.*, there will be a curve in the $\mathbf{s}$-space during a continuous time interval. If this $\mathbf{s}$-curve be referred to the $\mathbf{r}$-space, *i.e.*, if we determine the positions in $\mathbf{r}$-space of those points which have passed the fixed point $\mathbf{r}_1$, then these points make up a curve called a "streak line." It is given by

$$\mathbf{r} = \mathbf{F}(\mathbf{s}, t),$$

where

$$\mathbf{s} = \mathbf{G}(t, \mathbf{r}_1).$$

Therefore,

$$\mathbf{r} = \mathbf{F}\{\mathbf{G}(t, \mathbf{r}_1), t\}.$$

If the streak lines be required at time $t = t_1$ the first thing to do is to find from the equation $\mathbf{r} = \mathbf{F}(\mathbf{s}, t)$ the fluid particle $\mathbf{s}_1$ that has the coordinates of $\mathbf{r}_1$ at time $t = 0$ (this may be the time at which the motion begins), *i.e.*, moves past the point $\mathbf{r}_1$ at the instant $t = 0$. Then the point $\mathbf{r}$ must be found at which the fluid particle $\mathbf{s}_1 = \mathbf{G}(0, \mathbf{r}_1)$ is situated at time t_1, from the Lagrangian equation

$$\mathbf{r} = \mathbf{F}(\mathbf{s}_1, t_1) = \mathbf{F}\{\mathbf{G}(0, \mathbf{r}_1), t_1\}.$$

This gives one point on the streak line.   In a similar way values s_i are determined, for a sufficiently large number of time points between $t = 0$ and $t = t_1$, which have had the coordinates of r_1 at the appropriate times, and, finally, their positions are found at the time $t = t_1$.   As can be seen, the construction of streak lines is extremely tedious and difficult, particularly because the Lagrangian method is necessary.

An instantaneous photograph of the smoke line issuing from a chimney in the wind may be taken as an example of a streak line (this example is valid only in so far as the effects of the heat from the chimney on the air motion can be neglected).   Streak lines are occasionally used in the experimental investigation of the phenomena of flow of fluids.   This is done by introducing some coloring matter into the fluid through one or more small tubes and thereby, so to speak, dyeing the separate fluid particles. The lines of colored fluid thus made visible are streak lines. Their characteristic is obviously that every fluid particle running along a streak line must have passed by the mouth of the dye tube.   This method is also frequently used in the investigation of air flow, sometimes employing smoke.

Since, according to the above description, the streak lines are

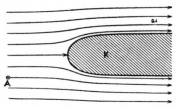

FIG.   44.—The   streamlines   of   a
steady flow.

forced from the end points, at the time being, of path lines, and since path lines are identical with streamlines in steady flow, the streak lines also coincide with the streamlines in steady flow.   Since experimental investigations using streak lines are generally made when the flow is steady, the streak lines give information about the streamlines and paths of particles.

37. Significance of System of Reference in Interpretation of the Form of Motion.—The choice of a coordinate system to which the fluid motion shall be referred is of great significance. As we shall see, it is possible in certain circumstances to change a non-steady flow into a steady one and *vice versa*.   We shall explain this by an example.

Consider the flow round a bridge pier with the system of reference—or, in other words, the observer—at rest, or consider the flow past the nose of an airship as it appears to an observer inside.   Both are examples of steady-state motions.   Figure 44

shows the path lines which in this case are coincident with the streak lines and the streamlines. It is to be noted that the body is assumed to be so long that back action produced on the flow round the front by disturbances at the back end can be neglected. If some dye (or smoke in the case of air) be introduced into the fluid at the point A, the colored line will lie along a streamline.

Matters are quite different however if the flow is not steady. This happens for example if the flow round an airship is referred

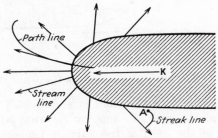

FIG. 45.—Streamlines, path lines, streak lines of a non-steady flow.

to a coordinate system at rest with respect to the undisturbed fluid. The streamlines, path lines, and streak lines then have forms which differ completely (Fig. 45). The body displaces the fluid particles in such a way that the particles in front are pushed forward, while the particles on each side are pushed simultaneously forward and sideways. The streamline diagram in this case is carried along with the body.

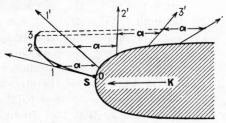

FIG. 46.—Construction of a path line.

38. Construction of Path Lines and Streak Lines.—As an example of the construction of a path line we shall consider the fluid particle **s** shown in Fig. 46. It is assumed that the body moves with uniform velocity, so that while the body K is displaced by an amount a in a small interval of time, the particle **s** moves approximately to the point 1 during this time, corresponding to

the magnitude of the velocity at the point 0. Since the stream-line diagram moves with the body, the particle **s** at the end of the interval of time will find itself on a new streamline 1′ which is separated from the previous one by an amount a in the direction of motion. In another small time interval the particle will move

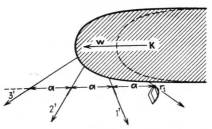

Fig. 47.—Construction of a streak line.

to 2 where it will come to a new streamline 2′, which is again separated from the previous one by distance a. In this way each element of the path of the particle becomes smaller than the previous one, so that finally the particle practically comes to rest.

We shall now consider the construction of a streak line (Fig. 47). To determine the shape of a streak line at the time $t = t_1$

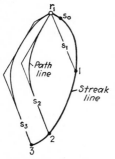

Fig. 48.—Details of the construction of a streak line.

we have to find the end points at that time of the path lines of all those fluid particles **s** which previously, *i.e.*, at $t \leqslant t_1$, moved through the point r_1 fixed in space. At the instant $t = t_1$, the particle s_0 will move through r_1 (Fig. 48). If, for simplicity, we again suppose that the body moves with uniform velocity, then at the time $t_1 - \Delta t$ the body was in the position shown by the dotted line in Fig. 47 and the streamline diagram therefore was displaced to the right by an amount a, so that the streamline 1′ passed through r_1. The fluid particle s_1 which at this instant moved through r_1 passed to the point 1 along the direction of the streamline 1′ during the interval Δt. At the instant $t_1 - 2\Delta t$, the fluid particle s_2 moved through r_1, and up to the time $t_1 - \Delta t$ it had described a path along the streamline 2′; further between $t_1 - \Delta t$ and t_1 it went back along 1′ to the point 2. All other particles behave in a similar manner. The end points of the

path lines thus constructed have all in the course of time passed through r_1; the line joining these points gives the required streak line through r_1.

39. Stream Tubes.—In conclusion we shall now introduce the concept of stream tube, which is very useful in the derivation of the simple laws of fluid motion.

If all the streamlines passing through a closed curve be drawn, they will, on account of the continuity of the velocity field, form a tube which we call a "stream tube." Since the streamlines everywhere have the direction of the velocity, this stream tube behaves like a tube with rigid walls inside which fluid flows. In general the shape of the tube will change from moment to moment, so that it is made up of ever changing fluid particles. But when the flow is steady so that the state of the fluid at any point is always the same, the stream tube behaves like a real tube. The content of a stream tube is also called "current filament."

CHAPTER VI

GEOMETRY OF THE VECTOR FIELD

40. Linear Vector Function of Position.—Let A be a fixed point in space with the coordinates x_0, y_0, z_0 and let it be assumed that the velocity in the fluid region considered is a regular function of the position. Let the velocity at the point A be

$$\mathbf{w}_0 = \mathbf{i}u_0 + \mathbf{j}v_0 + \mathbf{k}w_0,$$

where $\mathbf{i}$, $\mathbf{j}$, $\mathbf{k}$ are unit vectors in the x-, y-, z-directions. Then the velocity $\mathbf{w}$ in the neighborhood of A can be represented by the Taylor series:[1]

$$\mathbf{w} = \mathbf{w}_0 + \frac{\mathbf{r}_1 - \mathbf{r}_2}{1!}\circ\nabla\mathbf{w} + \frac{(\mathbf{r}_1 - \mathbf{r}_2)^2}{2!}\circ\nabla\nabla\mathbf{w} + \cdots,$$

where ∇ (pronounced "nabla") represents the vector operator:

$$\mathbf{i}\frac{\partial}{\partial x} + \mathbf{j}\frac{\partial}{\partial y} + \mathbf{k}\frac{\partial}{\partial z}.$$

Or in Cartesian coordinates:

$$u = u_0 + (x - x_0)\frac{\partial u}{\partial x} + (y - y_0)\frac{\partial u}{\partial y} + (z - z_0)\frac{\partial u}{\partial z} + \frac{(x - x_0)^2}{2!}\frac{\partial^2 u}{\partial x^2} + \cdots,$$

$$v = v_0 + (x - x_0)\frac{\partial v}{\partial x} + (y - y_0)\frac{\partial v}{\partial y} + (z - z_0)\frac{\partial v}{\partial z} + \frac{(x - x_0)^2}{2!}\frac{\partial^2 v}{\partial x^2} + \cdots,$$

$$w = w_0 + (x - x_0)\frac{\partial w}{\partial x} + (y - y_0)\frac{\partial w}{\partial y} + (z - z_0)\frac{\partial w}{\partial z} + \frac{(x - x_0)^2}{2!}\frac{\partial^2 w}{\partial x^2} + \cdots.$$

If we limit the discussion to the immediate neighborhood of the point A, the second order and following terms of the expansion can be neglected provided we make the region round A small enough. We have then in $\mathbf{w}$ a linear vector function of the

[1] The symbol ∘ represents a scalar vector operation, so that $\mathbf{a}\circ\mathbf{b}$ means the scalar product of the two vectors $\mathbf{a}$ and $\mathbf{b}$.

position, *i.e.*, there is always a region round A in which the velocity can be assumed to be linearly dependent on the distance from A without the error exceeding a negligibly small quantity. Such velocity fields in which the components of the velocity of a point are a linear function of its coordinates are called "fields of homogeneous deformation."

As the Taylor expansion shows, the difference between the velocity inside the region round A and the velocity at the point A is characterized by nine quantities, namely, the partial differentials of the three velocity components u, v, w, with respect to the three directions x, y, z:

$$\left.\begin{cases} \dfrac{\partial u}{\partial x} & \dfrac{\partial v}{\partial x} & \dfrac{\partial w}{\partial x} \\[2mm] \dfrac{\partial u}{\partial y} & \dfrac{\partial v}{\partial y} & \dfrac{\partial w}{\partial y} \\[2mm] \dfrac{\partial u}{\partial z} & \dfrac{\partial v}{\partial z} & \dfrac{\partial w}{\partial z} \end{cases}\right\}. \tag{1}$$

The generality of the treatment is in no wise limited if by a choice of the system of reference the point A is taken as the origin of the coordinate system and its velocity is taken as zero. The velocity at every point of the region round A will then be determined by the above nine quantities.

41. Geometrical Significance of the Individual Quantities of a Matrix Characterizing a Velocity Field.—What representation can we now make of the individual quantities in the above so-called "matrix" as given in

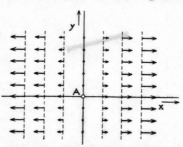

Fig. 49.—Velocity of expansion in the x-direction represented by

$$\mathbf{w} = x\mathbf{i}\frac{\partial u}{\partial x} = r\mathbf{oi}\,\mathbf{i}\frac{\partial u}{\partial x}.$$

Eq. (1)? Taking a simple case, first, let us suppose that all the terms of the matrix are zero except $\partial u/\partial x$, which we assume is positive.

Then the Taylor expansion given above reduces to

$$\mathbf{w} = x\mathbf{i}\frac{\partial u}{\partial x}$$

or, in scalar notation,

$$= x\frac{\partial u}{\partial x}, \qquad v = 0, \qquad w = 0.$$

Since the equation for the planes parallel to the (y,z)-plane is $x =$ constant, the velocity in these planes is constant, proportional to the distance of the plane from the point A and is in a direction parallel to the x-axis (Fig. 49). Since, moreover, $x =$ const. can be written $\mathbf{r} \circ \mathbf{i} =$ const., we have as an expression for the velocity field or, as we can also put it, for the displacement of these planes per unit time,

$$\mathbf{w} = x\mathbf{i}\frac{\partial u}{\partial x} = \mathbf{r} \circ \mathbf{i}\ \mathbf{i}\frac{\partial u}{\partial x}.$$

The analogous expressions, obtained by equating to zero all terms in the matrix except $\partial v/\partial y$ or $\partial w/\partial z$ are:

$$\mathbf{w} = y\mathbf{j}\frac{\partial v}{\partial y} = \mathbf{r} \circ \mathbf{j}\ \mathbf{j}\frac{\partial v}{\partial y}$$

and

$$\mathbf{w} = z\mathbf{k}\frac{\partial w}{\partial z} = \mathbf{r} \circ \mathbf{k}\ \mathbf{k}\frac{\partial w}{\partial z}.$$

The expression $\mathbf{w} = y\mathbf{j}\dfrac{\partial v}{\partial y} = \mathbf{r} \circ \mathbf{j}\ \mathbf{j}\dfrac{\partial v}{\partial y}$, which corresponds to a distribution in which the velocity is constant in planes parallel to the (z,x)-direction, is in a direction parallel to the y-axis and has a magnitude proportional to the distance of the plane from the point A.

If

$$\mathbf{w} = z\mathbf{k}\frac{\partial w}{\partial z} = \mathbf{r} \circ \mathbf{k}\ \mathbf{k}\frac{\partial w}{\partial z},$$

then the field is one in which the velocity is constant in planes parallel to the (x,y)-direction, is parallel to the z-axis, and has a magnitude proportional to the distance of the plane from the point A. The three expressions therefore represent velocities producing extension along the x-, y-, z-directions, respectively.

We obtain another kind of velocity distribution if all terms in the matrix are assumed to be zero except $\partial u/\partial y$.

Then

$$\mathbf{w} = y\mathbf{i}\frac{\partial u}{\partial y}$$

or

$$u = y\frac{\partial u}{\partial y}, \qquad v = 0, \qquad w = 0,$$

i.e., the velocity is constant in any plane parallel to the (z,x)-plane and increases proportional to the distance between that plane and the (z,x)-plane (Fig. 50). Since, moreover, we can write

$$\mathbf{r} \circ \mathbf{j} = \text{const.}$$

for $y = \text{const.}$, this velocity field can also be expressed in the form

$$\mathbf{w} = y\mathbf{i}\frac{\partial u}{\partial y} = \mathbf{r} \circ \mathbf{j}\, \mathbf{i}\frac{\partial u}{\partial y}.$$

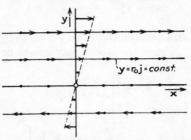

Corresponding velocity fields with similarly formed expressions can be obtained if each of the terms in the matrix is in turn assumed to differ from zero. If, for example, all the terms except $\partial v/\partial x$ disappear, the field will

Fig. 50.—Shearing velocity in the x-direction represented by
$$\mathbf{w} = y\mathbf{i}\frac{\partial u}{\partial y} = \mathbf{r} \circ \mathbf{j}\, \mathbf{i}\frac{\partial u}{\partial y}.$$

have velocities which are constant in planes parallel to the (y,z)-plane, proportional to their distance from the (y,z)-plane and directed parallel to the y-axis (Fig. 51).

With these velocity fields, or—as we may also say—with these changes in deformation per unit time, we are able to build up the most general (homogeneous) velocity fields or changes in

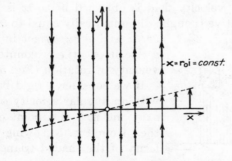

Fig. 51.—Shearing velocity in the y-direction represented by
$$\mathbf{w} = x\mathbf{j}\frac{\partial v}{\partial x} = \mathbf{r} \circ \mathbf{i}\, \mathbf{j}\frac{\partial v}{\partial x}.$$

deformation by superposition (which is permissible because of the linear dependence on the coordinates).

42. Shearing and Rotating Velocities.—In accordance with our principle of proceeding from the simple to the complicated, we shall now consider two particularly important cases which are

obtained by the superposition of two of the above velocity fields. First take the linear combination,

$$\mathbf{w} = x\mathbf{j}\frac{\partial v}{\partial x} + y\mathbf{i}\frac{\partial u}{\partial y} = \mathbf{r}\circ\mathbf{i}\,\mathbf{j}\frac{\partial v}{\partial x} + \mathbf{r}\circ\mathbf{j}\,\mathbf{i}\frac{\partial u}{\partial y} = \mathbf{r}\circ\left(\mathbf{ij}\frac{\partial v}{\partial x} + \mathbf{ji}\frac{\partial u}{\partial y}\right),$$

and assume that $\partial v/\partial x$ and $\partial u/\partial y$ are equal (positive or negative). From the above explanations and from Fig. 52 it is seen that

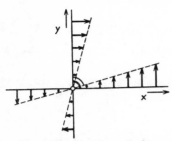

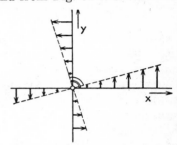

FIG. 52.—Shearing velocity
$$\mathbf{w} = x\mathbf{j}\frac{\partial v}{\partial x} + y\mathbf{i}\frac{\partial u}{\partial y} = \mathbf{r}\circ\left(\mathbf{ij}\frac{\partial v}{\partial x} + \mathbf{ji}\frac{\partial u}{\partial y}\right).$$

FIG. 53.—Rotation.

we are dealing with a change from what was originally a right angle to either an acute or an obtuse angle. This type of flow is known as a shearing motion.

A different velocity field is obtained if $\partial v/\partial x$ is positive but $\partial u/\partial y$ is negative though still numerically equal to $\partial v/\partial x$. Both $\partial v/\partial x$ and $-\partial u/\partial y$ act in the sense of a positive (*i.e.*, counter-clockwise) velocity of rotation (Fig. 53).

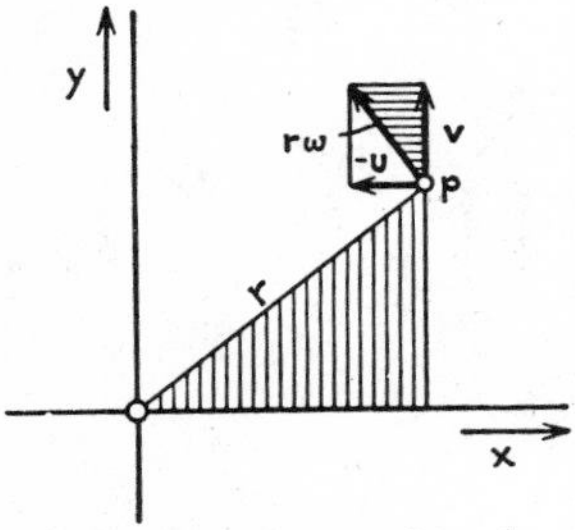

If we consider a rigid body rotating with an angular velocity ω about the z-axis, and if P is a point on the body, then from Fig. 54, owing to the similarity of the shaded triangles, we have

$$v:-u:r\omega = x:y:r,$$

FIG. 54.—Connection between mean rotation and angular velocity.

so that

$$u = -y\omega, \qquad v = x\omega, \qquad w = 0.$$

In this case $\partial v/\partial x = \omega$ and $\partial u/\partial y = -\omega$; the seven other differentials of the matrix are zero so that there is complete agreement with the case shown by Fig. 53.

We shall now move on from these two special cases to a somewhat more general one in which $\partial u/\partial y$ and $\partial v/\partial x$ have different

magnitudes. The shearing velocity will then be measured by the algebraic sum

$$\frac{\partial u}{\partial y} + \frac{\partial v}{\partial x}.$$

The mean rotation will be given by taking the mean of the two angular velocities $\partial v/\partial x$ and $-\partial u/\partial y$, which is

$$\frac{1}{2}\left(\frac{\partial v}{\partial x} - \frac{\partial u}{\partial y}\right).$$

[For the above rigid rotation this becomes $\frac{1}{2}\{\omega - (-\omega)\} = \omega$.]

43. The Concept of the Tensor.—We can construct fields of ever increasing generality by assuming fewer and fewer of the terms of the matrix (1) to be equal to zero. The most general (homogeneous) velocity field is clearly given by

$$\mathbf{w} = \mathbf{r} \circ \left\{ \mathbf{ii}\frac{\partial u}{\partial x} + \mathbf{ij}\frac{\partial v}{\partial x} + \mathbf{ik}\frac{\partial w}{\partial x} \right.$$
$$+ \mathbf{ji}\frac{\partial u}{\partial y} + \mathbf{jj}\frac{\partial v}{\partial y} + \mathbf{jk}\frac{\partial w}{\partial y}$$
$$\left. + \mathbf{ki}\frac{\partial u}{\partial z} + \mathbf{kj}\frac{\partial v}{\partial z} + \mathbf{kk}\frac{\partial w}{\partial z} \right\}, \qquad (2)$$

or

$$\mathbf{w} = \mathbf{r} \circ \left\{ \mathbf{i}\left(\mathbf{i}\frac{\partial u}{\partial x} + \mathbf{j}\frac{\partial v}{\partial x} + \mathbf{k}\frac{\partial w}{\partial x}\right) + \mathbf{j}\left(\mathbf{i}\frac{\partial u}{\partial y} + \mathbf{j}\frac{\partial v}{\partial y} + \mathbf{k}\frac{\partial w}{\partial y}\right) \right.$$
$$\left. + \mathbf{k}\left(\mathbf{i}\frac{\partial u}{\partial z} + \mathbf{j}\frac{\partial v}{\partial z} + \mathbf{k}\frac{\partial w}{\partial z}\right) \right\}.$$

Since $\mathbf{i}u + \mathbf{j}v + \mathbf{k}w = \mathbf{w}$, we can write

$$\frac{\partial \mathbf{w}}{\partial x} = \mathbf{i}\frac{\partial u}{\partial x} + \mathbf{j}\frac{\partial v}{\partial x} + \mathbf{k}\frac{\partial w}{\partial x}.$$

Corresponding expressions hold for $\partial \mathbf{w}/\partial y$ and $\partial \mathbf{w}/\partial z$, so that we can also write

$$\mathbf{w} = \mathbf{r} \circ \left(\mathbf{i}\frac{\partial \mathbf{w}}{\partial x} + \mathbf{j}\frac{\partial \mathbf{w}}{\partial y} + \mathbf{k}\frac{\partial \mathbf{w}}{\partial z}\right).$$

The quantities

$$\mathbf{i}\frac{\partial \mathbf{w}}{\partial x}, \qquad \mathbf{j}\frac{\partial \mathbf{w}}{\partial y}, \qquad \mathbf{k}\frac{\partial \mathbf{w}}{\partial z}$$

are known as "dyads." The rules governing these quantities

are the same as those of ordinary numbers except that dyads do not obey the law of commutative multiplication

$$\mathbf{i}\frac{\partial \mathbf{w}}{\partial x} \neq \frac{\partial \mathbf{w}}{\partial x}\mathbf{i}.$$

We can go yet a step farther toward the most general homogeneous field by writing

$$\mathbf{i}\frac{\partial \mathbf{w}}{\partial x} + \mathbf{j}\frac{\partial \mathbf{w}}{\partial y} + \mathbf{k}\frac{\partial \mathbf{w}}{\partial z} = \Phi.$$

This quantity Φ which, following Gibbs, we call a "tensor" is therefore an entity formed from three vectors $\partial \mathbf{w}/\partial x$, $\partial \mathbf{w}/\partial y$, $\partial \mathbf{w}/\partial z$, in the same way as a vector is formed from three scalars. The most general linear velocity field is therefore characterized by

$$\mathbf{w} = \mathbf{r} \circ \Phi.$$

Since the tensor has the special significance of completely determining a linear velocity field we shall now discuss some of the properties and peculiarities of this quantity.

44. Splitting a Tensor into a Symmetrical and Antisymmetrical Part.—If rows and columns are interchanged in the matrix (1) of Art. 40, we get

$$\frac{\partial \mathbf{w}}{\partial x}\mathbf{i} + \frac{\partial \mathbf{w}}{\partial y}\mathbf{j} + \frac{\partial \mathbf{w}}{\partial z}\mathbf{k} = \Phi_c.$$

The tensor Φ_c is said to be conjugate to Φ.

If in a special case the terms connected by the arrows are equal to each other,

$$\left\{ \begin{matrix} \dfrac{\partial u}{\partial x} & \dfrac{\partial v}{\partial x} & \dfrac{\partial w}{\partial x} \\[2mm] \dfrac{\partial u}{\partial y} & \dfrac{\partial v}{\partial y} & \dfrac{\partial w}{\partial y} \\[2mm] \dfrac{\partial u}{\partial z} & \dfrac{\partial v}{\partial z} & \dfrac{\partial w}{\partial z} \end{matrix} \right\},$$

so that

$$\frac{\partial u}{\partial y} = \frac{\partial v}{\partial x}, \qquad \frac{\partial u}{\partial z} = \frac{\partial w}{\partial x}, \qquad \frac{\partial v}{\partial z} = \frac{\partial w}{\partial y},$$

the tensor is called a "symmetrical tensor" and is uniquely determined by six quantities; it will be seen that a symmetrical tensor is conjugate to itself $\Phi = \Phi_c$.

If a tensor has the form

$$\begin{Bmatrix} 0 & \dfrac{\partial v}{\partial x} & \dfrac{\partial w}{\partial x} \\[2mm] \dfrac{\partial u}{\partial y} & 0 & \dfrac{\partial w}{\partial y} \\[2mm] \dfrac{\partial u}{\partial z} & \dfrac{\partial v}{\partial z} & 0 \end{Bmatrix}$$

with the terms connected by arrows being equal and opposite, *i.e.*,

$$\frac{\partial u}{\partial y} = -\frac{\partial v}{\partial x}, \quad \frac{\partial u}{\partial z} = -\frac{\partial w}{\partial x}, \quad \frac{\partial v}{\partial z} = -\frac{\partial w}{\partial y},$$

then the tensor is determined by three quantities and is said to be "antisymmetrical." In this case, $\Phi = -\Phi_c$.

It can now be shown that every tensor (characterized by nine terms) can be split up into a symmetrical part of six terms and an antisymmetrical part of three terms. The symmetrical part represents a velocity of pure extension along the three principal axes, while the antisymmetrical part represents a velocity of rotation. If Φ_c is the tensor conjugate to Φ then—as can be seen by carrying out the substitution—

$$\tfrac{1}{2}(\Phi + \Phi_c)$$

is the symmetrical part, and

$$\tfrac{1}{2}(\Phi - \Phi_c)$$

is the antisymmetrical part of the tensor.

In a velocity field characterized by a tensor, a spherical part of the fluid can transform only into an ellipsoid. Six quantities are necessary to determine this ellipsoid; three angles to fix the direction of the principal axes and three velocities of extension to determine their lengths. These make up the symmetrical part of the tensor. The three remaining terms are the components of a vector which determines the rotation of the fluid element, as will be seen in Art. 45.

As regards the relation of the tensor to the coordinate transformation it is to be noted that the components of the symmetric part transform like squares and products of point coordinates, while the components of the antisymmetric part transform like the coordinates, corresponding to the fact that it is a vector.

45. Stokes's Theorem.—We shall now discuss briefly the theorem of Stokes, and in Art. 46 that of Gauss; two important theorems in hydrodynamics.

Stokes's theorem deals with the concept of rotation and shows in Art. 47 that the vector determining the rotation is the anti-symmetric part of the tensor. Gauss's theorem introduces the new concept of divergence.

To derive Stokes's theorem we consider some closed curve **C** in a region in which we can neglect squares and higher powers in **r** in the Taylor series for **w**. In this region, then, the tensor Φ which uniquely determines the **w**-velocity field is practically constant.

If $d\mathbf{r}$ is a small element of **C**, the scalar product $\mathbf{w} \circ d\mathbf{r}$ is equal to the projection of the velocity at the point **r** of **C**, on to the

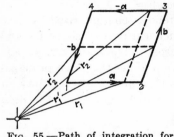

tangent to the curve **C** at this point. If the integral of this scalar product is taken round the closed curve **C** in a counter-clockwise direction, a so-called line integral is obtained

$$\oint^C \mathbf{w} \circ d\mathbf{r}.$$

If we substitute $\mathbf{w} = \mathbf{r} \circ \Phi$, we have

$$\oint^C \mathbf{r} \circ \Phi \circ d\mathbf{r}.$$

Fig. 55.—Path of integration for deriving Stokes's theorem.

We shall first take **C** to be the perimeter of a parallelogram 1, 2, 3, 4 (Fig. 55). It will be shown later that this does not limit the generality of the result. Since the direction of the path of integration from 1 to 2 is opposite to that from 3 to 4, these two parts contribute as follows:

$$\int_1^2 \mathbf{r}_1 \circ \Phi \circ d\mathbf{r} - \int_4^3 \mathbf{r}_2 \circ \Phi \circ d\mathbf{r} = (\mathbf{r}_1 - \mathbf{r}_2) \circ \Phi \circ \int_1^2 d\mathbf{r} = -\mathbf{b} \circ \Phi \circ \mathbf{a},$$

because

$$\mathbf{r}_1 - \mathbf{r}_2 = \text{const.} = -\mathbf{b}.$$

We get a corresponding expression for the parts of the line integral from 2 to 3 and from 4 to 1:

$$\int_2^3 \mathbf{r}_1' \circ \Phi \circ d\mathbf{r} - \int_1^4 \mathbf{r}_2' \circ \Phi \circ d\mathbf{r} = (\mathbf{r}_1' - \mathbf{r}_2') \circ \Phi \circ \int_2^3 d\mathbf{r} = \mathbf{a} \circ \Phi \circ \mathbf{b},$$

since

$$\mathbf{r}_1' - \mathbf{r}_2' = \text{const.} = \mathbf{a}.$$

Thus the complete line integral is

$$\oint^{1,2,3,4,1} \mathbf{r} \circ \Phi \circ d\mathbf{r} = \mathbf{a} \circ \Phi \circ \mathbf{b} - \mathbf{b} \circ \Phi \circ \mathbf{a}.$$

In Art. 43, we found for Φ the expression.

$$\Phi = \mathbf{i}\frac{\partial \mathbf{w}}{\partial x} + \mathbf{j}\frac{\partial \mathbf{w}}{\partial y} + \mathbf{k}\frac{\partial \mathbf{w}}{\partial z}.$$

Substituting only the first term $\Phi_1 = \mathbf{i}\dfrac{\partial \mathbf{w}}{\partial x}$, into the line integral, we find

$$\oint^{1,2,3,4} \mathbf{r} \circ \Phi_1 \circ d\mathbf{r} = \mathbf{a} \times \mathbf{b} \circ \mathbf{i} \times \frac{\partial \mathbf{w}}{\partial x}.[1]$$

Then with all three terms of the tensor,

$$\oint^{1,2,3,4,1} \mathbf{w} \circ d\mathbf{r} = \oint^{1,2,3,4,1} \mathbf{r} \circ \Phi \circ d\mathbf{r} =$$

$$\mathbf{a} \times \mathbf{b} \circ \left(\mathbf{i} \times \frac{\partial \mathbf{w}}{\partial x} + \mathbf{j} \times \frac{\partial \mathbf{w}}{\partial y} + \mathbf{k} \times \frac{\partial \mathbf{w}}{\partial z} \right).$$

$\mathbf{a} \times \mathbf{b} = \mathbf{A}$ is a vector whose direction is perpendicular to the surface $\mathbf{A}$ round which the line integral was taken and whose value is equal to the area of the surface.

$$\mathbf{i} \times \frac{\partial \mathbf{w}}{\partial x} + \mathbf{j} \times \frac{\partial \mathbf{w}}{\partial y} + \mathbf{k} \times \frac{\partial \mathbf{w}}{\partial z} = \mathbf{i}\left(\frac{\partial w}{\partial y} - \frac{\partial v}{\partial z} \right) + \mathbf{j}\left(\frac{\partial u}{\partial z} - \frac{\partial w}{\partial x} \right)$$
$$+ \mathbf{k}\left(\frac{\partial v}{\partial x} - \frac{\partial u}{\partial y} \right)$$

is a vector, which as a result of the geometrical meaning of the line integral, is a measure of the velocity of rotation. We represent it by $\mathbf{R}$ or rot $\mathbf{w}$ and call it the rotation of $\mathbf{w}$. As the derivation shows, the rotation is independent of the coordinate system. If instead of a parallelogram we take any other closed curve $\mathbf{C}$, it can easily be shown that the result derived above still holds good. To this end we subdivide the surface into a network of small parallelograms (Fig. 56), and in integrating round all these parallelograms all the line integrals will be zero (because each integration path is traversed in two opposite directions)

---

[1] If $\mathbf{a}, \mathbf{b}, \mathbf{c}, \mathbf{d}$ are four vectors, then

$$\mathbf{a} \times \mathbf{b} \circ \mathbf{c} \times \mathbf{d} = \mathbf{a} \circ \mathbf{c}\, \mathbf{b} \circ \mathbf{d} - \mathbf{b} \circ \mathbf{c}\, \mathbf{a} \circ \mathbf{d}$$
$$= \mathbf{a} \circ \mathbf{c}\, \mathbf{d} \circ \mathbf{b} - \mathbf{b} \circ \mathbf{c}\, \mathbf{d} \circ \mathbf{a}.$$

The symbol $\times$ represents a vector product, so that $\mathbf{a} \times \mathbf{b}$ means the product of the absolute values of $\mathbf{a}$ and $\mathbf{b}$ times the cos $(\mathbf{a}, \mathbf{b})$ *i.e.* $\mathbf{a} \times \mathbf{b} = a\, b \cos (\mathbf{a}, \mathbf{b})$.

except those round the outside edge of the parallelogram network. If the parallelograms be made smaller and smaller, then, in the limit as the sides of the parallelograms converge to zero, the integral round the outside edge of the network will become the integral round the closed curve **C**.

Therefore for every closed curve **C**, the line integral round **C** is equal to the scalar product of the surface vector **A** and the rotation **R**.

$$\oint^C \mathbf{w} \circ d\mathbf{r} = \mathbf{A} \circ \mathbf{R} = \mathbf{A} \circ \text{rot } \mathbf{w}. \tag{3}$$

As we have already remarked, the assumption that the velocity field in a small but finite region round the point A is a *linear* vector function is not entirely correct; it is an approximation which comes nearer and nearer the truth as the region decreases in size.

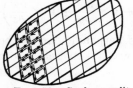

Fig. 56.—Surface split up into (infinitesimal) parallelograms.

Strictly speaking, the assumption of a linear dependence of the velocity vector on the position holds only for an infinitely small region, so that we should write

$$\lim_{dA\,=\,0} \oint \mathbf{w} \circ d\mathbf{r} = \lim_{dA\,=\,0} d\mathbf{A} \circ \mathbf{R} = \lim_{dA\,=\,0} d\mathbf{A} \circ \text{rot } \mathbf{w}. \tag{4}$$

Let **n** be a vector of unit length in the direction of the normal to the surface element dA, then $d\mathbf{A} = \mathbf{n}dA$ and

$$\text{rot } \mathbf{w} \circ \mathbf{n} = \mathbf{R} \circ \mathbf{n} = \lim_{dA\,=\,0} \frac{\oint \mathbf{w} \circ d\mathbf{r}}{dA}. \tag{3a}$$

$\mathbf{R} \circ \mathbf{n}$ is the component of **R** in the direction of **n**. **R** is therefore completely determined by the three rectangular components $\mathbf{n}_1$, $\mathbf{n}_2$, and $\mathbf{n}_3$. Equation $(3a)$ can therefore be used as a definition of **R** in a non-homogeneous velocity field. If **R** is chosen parallel to **n** so that $\mathbf{R} \circ \mathbf{n} = R$, then we have the following relation:

The rotation of a velocity field at a point A is equal to the limiting value to which the line integral of the velocity vector along a curve approximates, when the curve contains an infinitely small surface element perpendicular to the direction of rotation.

If we have a finite surface **A** (in which Φ and therefore **R** can no longer in general be regarded as constant), we can imagine it to be split up into infinitely small surface elements, for which

(4) is valid. On integrating round these infinitesimal areas the line integrals, whose paths lie completely inside the surface, again become zero, and only the integrals round the perimeter of the surface **A** are left. Thus

$$\oint^C \mathbf{w} \circ d\mathbf{r} = \iint^A \mathbf{R} \circ d\mathbf{A} = \iint^A \operatorname{rot} \mathbf{w} \circ d\mathbf{A}.$$

This is Stokes's theorem in its most general form; the line integral along **C** is transformed into a surface integral over **A** where **C** is the bounding contour of the surface **A**. As an important corollary of this theorem it should be noted that the right-hand integral $\iint^A \operatorname{rot} \mathbf{w} \circ d\mathbf{A}$ is zero where **A** is a closed surface, because the integration path **C** and therefore the left-hand integral becomes zero.

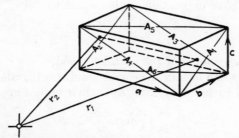

FIG. 57.—Parallelepiped for the derivation of Gauss's theorem. The symbols for the sides are written in the corresponding diagonals.

46. Gauss's Theorem.—We shall now investigate the integral analogous to $\oint^C \mathbf{w} \circ d\mathbf{r}$ for a closed surface **A**:

$$\oiint^A \mathbf{w} \circ d\mathbf{A} = \oiint^A \mathbf{r} \circ \Phi \circ d\mathbf{A}.$$

In this integral we again assume that in the region of **A** the velocity is a linear vector function, *i.e.*, Φ is constant. The geometrical significance of this integral is that it represents the volume displaced through the closed surface in unit time; in this case one speaks of a vector flux through the closed surface **A**.

We shall first assume that **A** is the surface of a rectangular parallelepiped with edges parallel to the coordinate axes.[1] Then,

[1] This assumption is not an absolute necessity; the argument can be carried through for a parallelepiped in any orientation, and although somewhat difficult, it can be shown that the definition given below for the divergence Θ is also independent of the coordinate system.

since $\mathbf{A}_1 = -\mathbf{A}_2, \quad \mathbf{A}_3 = -\mathbf{A}_4, \quad \mathbf{A}_5 = -\mathbf{A}_6$ (Fig. 57),

$$\oint^{\mathbf{A}_1} \mathbf{r}_1 \circ \Phi \circ d\mathbf{A} + \oint^{\mathbf{A}_2} \mathbf{r}_2 \circ \Phi \circ d\mathbf{A} = \oint^{\mathbf{A}_1} (\mathbf{r}_1 - \mathbf{r}_2) \circ \Phi \circ d\mathbf{A} = \mathbf{a} \circ \Phi \circ \mathbf{A}$$

$$= \mathbf{a} \circ \Phi \circ \mathbf{b} \times \mathbf{c},$$

since $\mathbf{A}_1 = \mathbf{b} \times \mathbf{c}$.

Substituting

$$\Phi = \mathbf{i}\frac{\partial \mathbf{w}}{\partial x} + \mathbf{j}\frac{\partial \mathbf{w}}{\partial y} + \mathbf{k}\frac{\partial \mathbf{w}}{\partial z},$$

and noting that $\mathbf{a}$ is parallel to $\mathbf{i}$ and perpendicular to $\mathbf{j}$ and $\mathbf{k}$ so that $\mathbf{a} \circ \mathbf{i} = a$, $\mathbf{a} \circ \mathbf{j} = \mathbf{a} \circ \mathbf{k} = 0$, we obtain

$$\mathbf{a} \circ \Phi \circ \mathbf{b} \times \mathbf{c} = a\frac{\partial \mathbf{w}}{\partial x} \circ \mathbf{b} \times \mathbf{c} = a\left(\mathbf{i}\frac{\partial u}{\partial x} + \mathbf{j}\frac{\partial v}{\partial x} + \mathbf{k}\frac{\partial \mathbf{w}}{\partial x}\right) \circ \mathbf{b} \times \mathbf{c}.$$

Moreover, since $\mathbf{j}$ and $\mathbf{k}$ both are perpendicular to $\mathbf{b} \times \mathbf{c}$ and $a\mathbf{i} = \mathbf{a}$, the above expression is also equal to

$$\frac{\partial u}{\partial x}\mathbf{a} \circ \mathbf{b} \times \mathbf{c} = \frac{\partial u}{\partial x}V,$$

where V is the volume of the parallelepiped.

Similarly we can obtain the corresponding integrals for the two other pairs of surfaces $\mathbf{A}_{3,4}$ and $\mathbf{A}_{5,6}$; their values are

$$\frac{\partial v}{\partial y}V, \text{ and } \frac{\partial w}{\partial z}V,$$

so that for the integral over the whole surface we get

$$\oint^{\mathbf{A}} \mathbf{w} \circ d\mathbf{A} = \oint^{\mathbf{A}} \mathbf{r} \circ \Phi \circ d\mathbf{A} = \left(\frac{\partial u}{\partial x} + \frac{\partial v}{\partial y} + \frac{\partial w}{\partial z}\right)V$$

$$= \left(\mathbf{i} \circ \frac{\partial \mathbf{w}}{\partial x} + \mathbf{j} \circ \frac{\partial \mathbf{w}}{\partial y} + \mathbf{k} \circ \frac{\partial \mathbf{w}}{\partial z}\right)V.$$

The expression

$$\mathbf{i} \circ \frac{\partial \mathbf{w}}{\partial x} + \mathbf{j} \circ \frac{\partial \mathbf{w}}{\partial y} + \mathbf{k} \circ \frac{\partial \mathbf{w}}{\partial z} = \frac{\partial u}{\partial x} + \frac{\partial v}{\partial y} + \frac{\partial w}{\partial z}$$

is a scalar and is called the divergence of the vector $\mathbf{w}$. It is written as Θ or as div $\mathbf{w}$. The divergence is therefore the volume pushed out through the boundary surface per unit time and per unit volume.

As in the derivation of Stokes's theorem, the results obtained for the parallelepiped can be applied to any shape of volume, so long as the velocity distribution in this volume can be approximated by a linear vector function. This derivation is only

strictly valid for an element of volume dV, so that the definition of the divergence of any velocity field at a point is

$$\text{div } \mathbf{w} = \Theta = \lim_{\Delta V = 0} \frac{\oint^{\mathbf{A}} \mathbf{w} \circ d\mathbf{A}}{\Delta V},$$

or in words:

The divergence at a point in a velocity field is the flux of the vector $\mathbf{w}$ per unit volume through the surface of an infinitely small volume element.

If we consider a finite region of fluid with a tensor dependent on the position, the last equation gives Gauss's relation, which transforms a surface integral into a volume integral

$$\oint^{\mathbf{A}} \mathbf{w} \circ d\mathbf{A} = \iiint^{V} \text{div } \mathbf{w} \, dV.$$

Gauss's theorem therefore states that the flux of the vector through a closed surface is equal to the volume integral of the divergence of this vector taken over the surface of the enclosed volume.

47. Introduction of the Operator ∇.—By using the Hamiltonian Nabla operator

$$\nabla = \mathbf{i}\frac{\partial}{\partial x} + \mathbf{j}\frac{\partial}{\partial y} + \mathbf{k}\frac{\partial}{\partial z},$$

which is to be treated as a vector, the tensor

$$\Phi = \mathbf{i}\frac{\partial \mathbf{w}}{\partial x} + \mathbf{j}\frac{\partial \mathbf{w}}{\partial y} + \mathbf{k}\frac{\partial \mathbf{w}}{\partial z}$$

can also be written

$$\Phi = \nabla \mathbf{w}.$$

The tensor conjugate to Φ, obtained by interchanging the rows and columns in the matrix, can obviously be written

$$\Phi_c = \frac{\partial \mathbf{w}}{\partial x}\mathbf{i} + \frac{\partial \mathbf{w}}{\partial y}\mathbf{j} + \frac{\partial \mathbf{w}}{\partial z}\mathbf{k} = \mathbf{w}\nabla.$$

Clearly

$$\nabla \mathbf{w} \equiv \tfrac{1}{2}(\nabla \mathbf{w} + \mathbf{w}\nabla) + \tfrac{1}{2}(\nabla \mathbf{w} - \mathbf{w}\nabla).$$

Also

$\tfrac{1}{2}(\nabla \mathbf{w} + \mathbf{w}\nabla)$ is the symmetric part of the tensor $\Phi = \nabla \mathbf{w}$ and $\tfrac{1}{2}(\nabla \mathbf{w} - \mathbf{w}\nabla)$ is the antisymmetric part of the tensor $\Phi = \nabla \mathbf{w}$, as can be seen immediately by writing the two tensors $\tfrac{1}{2}(\nabla \mathbf{w} + \mathbf{w}\nabla)$ and $\tfrac{1}{2}(\nabla \mathbf{w} - \mathbf{w}\nabla)$ in their matrix form.

We shall now introduce another method of writing the rotation **R** and the divergence Θ. In Art. 45 we found that the rotation **R** is expressed by

$$\operatorname{rot} \mathbf{w} = \mathbf{R} = \mathbf{i} \times \frac{\partial \mathbf{w}}{\partial x} + \mathbf{j} \times \frac{\partial \mathbf{w}}{\partial y} + \mathbf{k} \times \frac{\partial \mathbf{w}}{\partial z},$$

$$\operatorname{rot} \mathbf{w} = \mathbf{i}\frac{\partial}{\partial x} \times \mathbf{w} + \mathbf{j}\frac{\partial}{\partial y} \times \mathbf{w} + \mathbf{k}\frac{\partial}{\partial z} \times \mathbf{w} = \boldsymbol{\nabla} \times \mathbf{w}.$$

Or written out fully in coordinates:

$$\operatorname{rot} \mathbf{w} = \boldsymbol{\nabla} \times \mathbf{w} = \left(\mathbf{i}\frac{\partial}{\partial x} + \mathbf{j}\frac{\partial}{\partial y} + \mathbf{k}\frac{\partial}{\partial z}\right) \times (\mathbf{i}u + \mathbf{j}v + \mathbf{k}w)$$

$$= \mathbf{i}\left(\frac{\partial w}{\partial y} - \frac{\partial v}{\partial z}\right) + \mathbf{j}\left(\frac{\partial u}{\partial z} - \frac{\partial w}{\partial x}\right) + \mathbf{k}\left(\frac{\partial v}{\partial x} - \frac{\partial u}{\partial y}\right).$$

This can also be written in determinant form:

$$\operatorname{rot} \mathbf{w} = \begin{vmatrix} \mathbf{i} & \mathbf{j} & \mathbf{k} \\ \dfrac{\partial}{\partial x} & \dfrac{\partial}{\partial y} & \dfrac{\partial}{\partial z} \\ u & v & w \end{vmatrix}.$$

Similarly, for the divergence, we have

$$\operatorname{div} \mathbf{w} = \Theta = \mathbf{i}\circ\frac{\partial \mathbf{w}}{\partial x} + \mathbf{j}\circ\frac{\partial \mathbf{w}}{\partial y} + \mathbf{k}\circ\frac{\partial \mathbf{w}}{\partial z}$$

$$= \mathbf{i}\frac{\partial}{\partial x}\circ\mathbf{w} + \mathbf{j}\frac{\partial}{\partial y}\circ\mathbf{w} + \mathbf{k}\frac{\partial}{\partial z}\circ\mathbf{w}$$

$$= \boldsymbol{\nabla}\circ\mathbf{w};$$

or in coordinates

$$\operatorname{div} \mathbf{w} = \boldsymbol{\nabla}\circ\mathbf{w} = \left(\mathbf{i}\frac{\partial}{\partial x} + \mathbf{j}\frac{\partial}{\partial y} + \mathbf{k}\frac{\partial}{\partial z}\right)\circ(\mathbf{i}u + \mathbf{j}v + \mathbf{k}w)$$

$$= \frac{\partial u}{\partial x} + \frac{\partial v}{\partial y} + \frac{\partial w}{\partial z}.$$

Collecting these results we have:
Tensor,

$$\Phi = \boldsymbol{\nabla}\mathbf{w},$$

Vector,

$$\mathbf{R} = \boldsymbol{\nabla} \times \mathbf{w} = \operatorname{rot} \mathbf{w},$$

Scalar,

$$\Theta = \boldsymbol{\nabla}\circ\mathbf{w} = \operatorname{div} \mathbf{w}.$$

In conclusion we shall illustrate the utility of this new symbolic method by two examples:

1. Given a vector field whose vector **w** is itself a gradient of a scalar field φ,

$$\mathbf{w} = \text{grad } \varphi = \boldsymbol{\nabla}\varphi.$$

We first write down the divergence of this vector field

$$\text{div } \mathbf{w} = \text{div grad } \varphi = \boldsymbol{\nabla}{\circ}\boldsymbol{\nabla}\varphi = \boldsymbol{\nabla}^2\varphi = \frac{\partial^2\varphi}{\partial x^2} + \frac{\partial^2\varphi}{\partial y^2} + \frac{\partial^2\varphi}{\partial z^2} = \Delta\varphi,$$

where

$$\Delta = \frac{\partial^2}{\partial x^2} + \frac{\partial^2}{\partial y^2} + \frac{\partial^2}{\partial z^2}$$

is the Laplace operator.

If we write down the rotation we see at once that

$$\text{rot grad } \varphi = \boldsymbol{\nabla} \times \boldsymbol{\nabla}\varphi = 0,$$

since the vector product of a vector with itself is zero. (The same result can be obtained of course with coordinates in a somewhat more complicated way; expressions of the form

$$\frac{\partial^2\varphi}{\partial x\partial y} - \frac{\partial^2\varphi}{\partial y\partial x} = 0$$

are then obtained.)

2. If the vector field **w** is the rotation of another vector **A**, its divergence is

$$\text{div } \mathbf{w} = \text{div rot } \mathbf{A} = \boldsymbol{\nabla}{\circ}\boldsymbol{\nabla} \times \mathbf{A} = 0,$$

since $\boldsymbol{\nabla} \times \mathbf{A}$ represents a vector with a direction perpendicular to $\boldsymbol{\nabla}$ and the scalar multiplication of two perpendicular vectors is zero. The rotation[1] of the vector field **w** is

$$\begin{aligned}\text{rot } \mathbf{w} = \text{rot rot } \mathbf{A} &= \boldsymbol{\nabla} \times (\boldsymbol{\nabla} \times \mathbf{A})\\ &= \boldsymbol{\nabla}{\circ}(\mathbf{A}\boldsymbol{\nabla} - \boldsymbol{\nabla}\mathbf{A}) = \boldsymbol{\nabla}{\circ}\mathbf{A}\boldsymbol{\nabla} - \boldsymbol{\nabla}{\circ}\boldsymbol{\nabla}\mathbf{A}.\end{aligned}$$

Therefore

$$\text{rot } \mathbf{w} = \text{rot rot } \mathbf{A} = \text{grad div } \mathbf{A} - \Delta\mathbf{A}.$$

We have thus found that a vector field, formed by the gradient of a scalar, has no rotation (rotation-free field), while a vector field obtained by the rotation of a vector has no divergence (source-free field).

[1] If **a, b,** and **c** are three vectors, then

$$\begin{aligned}\mathbf{a} \times (\mathbf{b} \times \mathbf{c}) &= \mathbf{a}{\circ}(\mathbf{c}\mathbf{b} - \mathbf{b}\mathbf{c})\\ &= \mathbf{a}{\circ}\mathbf{c} \ \mathbf{b} - \mathbf{a}{\circ}\mathbf{b} \ \mathbf{c}.\end{aligned}$$

Inverse problems are: (1) to find a vector field whose rotation is everywhere zero and whose divergence is everywhere determined; and (2) to find a vector field whose divergence is everywhere zero and whose rotation is everywhere determined. Both can be solved by the use of Laplace's operator with the methods of potential theory. If a general field is given, it is always possible to split it up into two parts one with zero rotation and the other with zero divergence.

CHAPTER VII

ACCELERATION OF A FLUID PARTICLE

48. Velocity Change of a Fluid Particle as a Function of the Time and the Velocity Field.—The acceleration of a fluid particle is the rate of change in its velocity. As a prelude we shall discuss the more general problem of determining the rate of change with time of any scalar or vector (such as density, temperature, velocity, rotation, or deformation).

The problem can be approached in two ways, corresponding to the methods of Euler and Lagrange. One can ask either: How does the quantity concerned, *e.g.*, the velocity, vary at a given point $\mathbf{r}$ in the space filled by the fluid? or, How does the velocity alter of a given particle $\mathbf{s}$ in the fluid? In the first case (of a fixed point $\mathbf{r}$ in space) we speak of "local" differentials, in the second case (with our attention fixed on some definite particle $\mathbf{s}$) of "substantial" differentials. If we assume for example that the temperature T is dependent on time (and in general on position), Euler's method will give the local differentials $(\partial T/\partial t)_\mathbf{r}$, or merely $\partial T/\partial t$, since this is an ordinary partial differential.

To distinguish the substantial differential from the local differential of Euler, we shall write the substantial differential as DT/dt (read as "T substantial with respect to t").

49. The Substantial Differential Is the Sum of the Local and Convective Differentials.—We shall now see that the substantial differential can be split up into two parts. Let us find out how a certain property of a definite fluid particle—here we choose its temperature—varies with the time. At a given moment $t = t_1$, the fluid particle has the vector of position $\mathbf{r}_1$. Imagine that the temperature distribution of the space varies with the time; then a temperature alteration can be assigned to every position $\mathbf{r}_1$. If the fluid particle is at rest, this will be the *whole* fluctuation. But if the particle moves, then in general its temperature will be affected by its change in position, because even if at each point $\mathbf{r}_1$ the temperature is constant with time, the temperature

of a particle changes on account of its getting into a different position **r**. This latter part of the change is called a "convective" temperature alteration. The convective alteration depends on:

1. The direction of motion and the magnitude of the velocity of the particle, therefore on the velocity vector.

2. The temperature distribution in space or the "temperature gradient."

Consider a group of curves of T = constant as shown in Fig. 58. If the particle moves from $\mathbf{r}_1$ to $\mathbf{r}_2$ in the time dt, the tem-

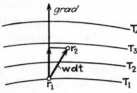

FIG. 58.—Motion of a fluid element from $\mathbf{r}_1$ to $\mathbf{r}_2$ in a temperature field which is a function of position. The convective change of temperature of this element is then $\mathbf{W} \circ$ grad T.

perature will change by an amount equal to the projection of the vector $\mathbf{w}dt$ on the normal at $\mathbf{r}_1$ multiplied by grad T; the temperature change is therefore equal to the scalar product of $\mathbf{w}dt$ and grad T. The convective change per unit time, *i.e.*, the convective differential, is therefore given by

$$\mathbf{w} \circ \text{grad } T.$$

If instead of the temperature we take any other quantity ω, which may be a scalar, a vector or even a tensor, then it is easy to prove that in this case also:

$$\frac{D\omega}{dt} = \frac{\partial \omega}{\partial t} + \mathbf{w} \circ \text{grad } \omega. \tag{1}$$

For the particularly important substantial differential of the velocity we therefore have

$$\frac{D\mathbf{w}}{dt} = \frac{\partial \mathbf{w}}{\partial t} + \mathbf{w} \circ \text{grad } \mathbf{w}. \tag{2}$$

By remembering that

$$\mathbf{w} \circ \text{grad } \mathbf{w} = \mathbf{w} \circ \nabla \mathbf{w} = (\mathbf{i}u + \mathbf{j}v + \mathbf{k}w) \circ \left(\mathbf{i}\frac{\partial \mathbf{w}}{\partial x} + \mathbf{j}\frac{\partial \mathbf{w}}{\partial y} + \mathbf{k}\frac{\partial \mathbf{w}}{\partial z} \right)$$

$$= u\frac{\partial \mathbf{w}}{\partial x} + v\frac{\partial \mathbf{w}}{\partial y} + w\frac{\partial \mathbf{w}}{\partial z}$$

$$= \mathbf{i}\left(u\frac{\partial u}{\partial x} + v\frac{\partial u}{\partial y} + w\frac{\partial u}{\partial z} \right)$$

$$+ \mathbf{j}\left(u\frac{\partial v}{\partial x} + v\frac{\partial v}{\partial y} + w\frac{\partial v}{\partial z} \right)$$

$$+ \mathbf{k}\left(u\frac{\partial w}{\partial x} + v\frac{\partial w}{\partial y} + w\frac{\partial w}{\partial z} \right),$$

Eq. (2), written in coordinates, becomes

$$\frac{\partial u}{\partial t} + u\frac{\partial u}{\partial x} + v\frac{\partial u}{\partial y} + w\frac{\partial u}{\partial z} \cdots x - \text{component}$$
$$\frac{\partial v}{\partial t} + u\frac{\partial v}{\partial x} + v\frac{\partial v}{\partial y} + w\frac{\partial v}{\partial z} \cdots y - \text{component} \Bigg\}. \quad (2a)$$
$$\frac{\partial w}{\partial t} + u\frac{\partial w}{\partial x} + v\frac{\partial w}{\partial y} + w\frac{\partial w}{\partial z} \cdots z - \text{component}$$

The Lagrangian expression for the substantial differential of the quantity **w** is simply

$$\left(\frac{\partial \mathbf{w}}{\partial t}\right)_s,$$

but the symbols for the local differential are correspondingly complicated since the vector **r** appears as a dependent quantity. For problems in one dimension the method of Lagrange can be used with advantage, as was shown by Riemann.[1]

50. Kinematic Boundary Conditions; Theorem of Lagrange.— The kinematic boundary conditions at the surface of contact between a liquid and a solid body, and also between two nonmiscible fluids (water and oil, water and air, etc.), must clearly be such that neither vacuum nor interpenetration can occur. The necessary consequence

FIG. 59.—A fluid element in the interior of the fluid cannot reach the surface in a finite time (Lagrange's theorem).

of this is that the normal components of the velocities of the two media are equal on each side of the surface of contact, *i.e.*,

$$w_{\mathbf{n}} = u \cos (\mathbf{n}, x) + v \cos (\mathbf{n}, y) + w \cos (\mathbf{n}, z)$$

must be equal on both sides of the surface of contact.

We shall now consider an important theorem due to Lagrange, which states that the boundary surface is always composed of the same particles. The law in this general form is not quite correct as we shall see later, but it can easily be proved that any fluid particle not at the surface will never reach the surface in a finite time. That particles on the surface can never penetrate the interior of the fluid is not true however.

Consider, as in Fig. 59, a point A at a small distance h from the surface. If h is small enough we can surround A by a region in which the deformation can be assumed to be linear, with sufficient

accuracy, so that for example straight lines in the region considered are transformed to straight lines, points of intersection to points of intersection, uniform divisions to uniform divisions, and so on. The tensor characterizing the deformation is therefore assumed constant in the region round A.

Let us suppose that the fluid particle at A moves up to the boundary, so that the decrease of h with time is proportional to h. We shall assume that the constant of proportion which may be a function of time always remains finite, therefore

$$\frac{dh}{dt} = g(t)h,$$

or if h_0 is a constant of integration,

$$\ln \frac{h}{h_0} = \int_0^t g(t)dt.$$

Since the integrand was assumed to be finite, the integral must also be finite in a finite time, *i.e.*, $\ln h/h_0$ and therefore also h must be finite. While the integral can be positive or negative, h must always be positive and finite. Although h decreases without limit with negative values of $g(t)$, it does not become zero in a finite time.

Thus it has been shown that a fluid particle not situated on the surface of a fluid at a given instant can never reach the surface in a finite time. It is of course assumed that the velocity field or the bounding surface shows no instability, since in this case the hypothesis of linear deformation for the domain under consideration is no longer tenable. If, for example, we consider the motion in the neighborhood of a sharp edge of zero angle, then the fluid is, as it were, split by the edge, and thus fluid particles from the interior reach the bounding surface.

51. Liquids and Gases Are Not to Be Considered as Ideal Media but as Quasi-continua.—The fact that fluid particles which belong to the boundary can still, under certain conditions, reach the interior of the liquid is connected with the fact that liquids and gases must be considered as quasi-continua and not as ideal media. If we consider the motion of a non-viscous fluid round a cylinder or sphere, then, as will be seen later, the streamlines are of the form shown in Fig. 60, and the velocity at the points A and B, where the streamlines cut the body, is zero. In the immediate neighborhood of these two points the velocity field can be represented by a constant tensor. At A small lines

at right angles to the surface are continuously becoming shorter, and similar lines parallel to the surface are continuously being elongated; at B the state of affairs is exactly reversed. A particle in the neighborhood of A, at a small distance from the sphere, approaches closer and closer to the bounding surface but never reaches it; the distance apart follows an exponential law; it never becomes zero, but it soon becomes smaller than any assignable small quantity.

On the other hand the number of particles at A, which were there originally, always decreases. Their number follows an

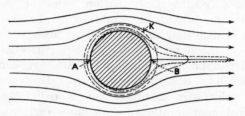

FIG. 60.—Flow round a cylinder in the first instant of motion from rest. The "fluid surface" K (— line) changes shape to the -·-·- form and later to the ...... form.

exponential law, so that after a short time only very few of the original fluid particles remain in the vicinity of the stagnation point A.

Another way of looking at this is as follows: Let a fluid surface be one which is constantly composed of the same fluid particles. Consider a spherical fluid surface very near the sphere, as is shown by the broken line K, Fig. 60. After a short interval of time this surface is in the position given by the dash-dot line and shortly afterward by the dotted line. In the front of the body the surface approaches it rapidly, while in the rear it becomes more and more elongated. All these conclusions are based on the assumption of the velocity field shown in Fig. 60, which, it might be noted, is not that of a real (viscous) fluid.

CHAPTER VIII

EQUATION OF CONTINUITY

52. Incompressible Homogeneous Fluids.—In Chap. VI we dealt with velocity fields of great generality, but we must now make a reservation and this is that not all vector distributions give rise to a possible fluid motion. One of the limitations follows from the important condition of the conservation of matter. It is true that so long as there are no laws about the dependence of the density of the fluid on space and time, any vector field can be made to fulfill the above condition by a correct choice of the density at each point. However, the density cannot be arbitrarily assumed but is definitely determined from physical considerations. Therefore certain vector fields will lie outside the scope of our investigation since they do not satisfy the important requirement that matter can be neither produced nor annihilated.

First let us consider the most important practical case of a density which is constant in space and in time, *i.e.*, an incompressible fluid. The condition for the conservation of matter then is satisfied, if the amount of fluid flowing out of a definite volume element equals the amount flowing into it. In other words, the divergence of the field has to be zero, for which in Art. 46 we found the analytical expression:

$$\nabla \circ \mathbf{w} = \operatorname{div} \mathbf{w} = 0$$

or

$$\frac{\partial u}{\partial x} + \frac{\partial v}{\partial y} + \frac{\partial w}{\partial z} = 0.$$

Equation of continuity for a homogeneous incompressible fluid.

For the steady-state case it is possible to express the continuity in another way by means of stream tubes. The same amount of fluid must flow through every cross section of a stream tube, which in this case guides the fluid as if it were a rigid tube. If the shape of the streamlines is known, as for example in pipes where approximately th ewhole pipe may be considered a stream tube, then for every section A of the tube we have

$$A_1 w_1 = A_2 w_2 = A w = \text{const.}$$

53. Eulerian Derivation of the Continuity Equation for Gases.
We shall now consider the general case, in which the mass density
ρ is an arbitrary analytic function of space and time. The equation of continuity can be derived either by considering a fixed
volume ΔV in the **r**-space (method of Euler) or by considering
some definite fluid mass in the **s**-space (method of Lagrange).
In this article the first method will be followed.

The increase or decrease of the fluid mass in the volume ΔV
is an expression involving the excess or deficiency of inflow over
outflow as given by the integral

$$\oint^{\mathbf{A}} \rho \mathbf{w} \circ d\mathbf{A},$$

where $\mathbf{A}$ is the area of the boundary surface of the volume and
the positive direction of the normal is taken outward. This
integral is negative with an increase of mass and positive with
a decrease. We transform this surface integral by Gauss's
theorem (Art. 46) into a volume integral and obtain the expression

$$\int^{\Delta V} \text{div } (\rho \mathbf{w}) dV.$$

Since we can approximate any vector field by a linear vector
system if only the volume be made small enough, the divergence
can be considered constant in the small volume ΔV so that

$$\oint^{\mathbf{A}} \rho \mathbf{w} \circ d\mathbf{A} = \Delta V \text{ div } (\rho \mathbf{w}).$$

On account of the conservation of matter this expression must
be equal to the increase or decrease per unit time of the mass
in the constant volume ΔV. Therefore

$$-\Delta V \text{ div } (\rho \mathbf{w}) = \Delta V \frac{\partial \rho}{\partial t}$$

or

$$\frac{\partial \rho}{\partial t} + \text{div } (\rho \mathbf{w}) = 0. \tag{1}$$

This general equation of continuity can be given a somewhat
different form. On account of

$$\text{div } (\rho \mathbf{w}) = \mathbf{\nabla} \circ (\rho \mathbf{w}) = \mathbf{\nabla} \rho \circ \mathbf{w} + \rho \mathbf{\nabla} \circ \mathbf{w} = \mathbf{w} \circ \text{grad } \rho + \rho \text{ div } \mathbf{w},$$

we have

$$\frac{\partial \rho}{\partial t} + \mathbf{w} \circ \text{grad } \rho + \rho \text{ div } \mathbf{w} = 0; \tag{1a}$$

or in Cartesians

$$\frac{\partial \rho}{\partial t} + u\frac{\partial \rho}{\partial x} + v\frac{\partial \rho}{\partial y} + w\frac{\partial \rho}{\partial z} + \rho\left(\frac{\partial u}{\partial x} + \frac{\partial v}{\partial y} + \frac{\partial w}{\partial z}\right) = 0. \quad (1b)$$

If we derive the equation of continuity for a definite mass of fluid, then the equation for the conservation of matter is clearly

$$\rho\Delta V = \text{const.}$$

Therefore, since we are dealing with the time change of a definite fluid mass, *i.e.*, with substantial differentiation:

$$\frac{D}{dt}(\rho\Delta V) = 0$$

or

$$\rho\frac{D}{dt}\Delta V + \Delta V\frac{D\rho}{dt} = 0.$$

But, since $D\Delta V/dt$ is the change in the volume ΔV per unit time, and div **w** the change per unit volume, we have

$$\frac{D}{dt}\Delta V = \Delta V \text{ div } \mathbf{w}.$$

Further, since from page 96

$$\frac{D\rho}{dt} = \frac{\partial \rho}{\partial t} + \mathbf{w}\circ \text{grad } \rho,$$

we get on dividing by ΔV

$$\rho \text{ div } \mathbf{w} + \frac{\partial \rho}{\partial t} + \mathbf{w}\circ \text{grad } \rho = 0,$$

which is our Eq. (1a). If we are dealing with a steady motion, ρ is independent of t and the equation of continuity becomes

$$\text{div } (\rho\mathbf{w}) = \mathbf{w}\circ \text{grad } \rho + \rho \text{ div } \mathbf{w} = 0. \quad (2)$$

If in addition ρ is constant over space, ρ div **w** = 0, and therefore

$$\text{div } \mathbf{w} = 0. \quad (3)$$

If we consider the motion through a stream tube of varying cross section A, then in order that the mass flow through the surface of each tube be zero, we must have

$$A_1 w_1 \rho_1 = A_2 w_2 \rho_2 = \text{const.} = A w \rho.$$

This equation has a meaning only when we know the form of the streamlines, as in the case of motion through pipes.

For fluids of uniform density (ρ = constant) this method gives the form of the equation of continuity which is familiar in hydraulics, namely,

$$Aw = \text{const.}$$

54. The General Lagrangian Equation of Continuity.—We shall now derive the equation of continuity by adopting the Lagrangian method.

If dV is a definite volume element in the **s**-space with sides of length da, db, dc (Fig. 61) and density ρ_0, then the mass of this fluid element is $\rho_0 dadbdc$. If we investigate the shape of this volume of fluid dV in the **r**-space at some later instant of

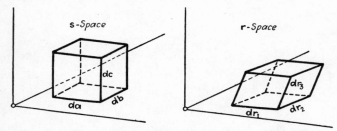

Fig. 61.—Corresponding volume elements in **r**- and in **s**-space.

time t, then, if $d\mathbf{r}_1$, $d\mathbf{r}_2$, $d\mathbf{r}_3$ denote the sides of the deformed volume element,

$$d\mathbf{r}_1 = \frac{\partial \mathbf{r}}{\partial a}da,$$

$$d\mathbf{r}_2 = \frac{\partial \mathbf{r}}{\partial b}db,$$

$$d\mathbf{r}_3 = \frac{\partial \mathbf{r}}{\partial c}dc.$$

The volume of this deformed element is now

$$d\mathbf{r}_1 \circ d\mathbf{r}_2 \times d\mathbf{r}_3,$$

and, if the density of the deformed volume element is equal to ρ, the condition of conservation of matter becomes

$$\rho_0 dadbdc = \rho d\mathbf{r}_1 \circ d\mathbf{r}_2 \times d\mathbf{r}_3$$

or

$$\rho_0 dadbdc = \rho dadbdc \frac{\partial \mathbf{r}}{\partial a} \circ \frac{\partial \mathbf{r}}{\partial b} \times \frac{\partial \mathbf{r}}{\partial c}.$$

The Lagrangian equation of continuity therefore is

$$\rho_0 = \rho \frac{\partial \mathbf{r}}{\partial a} \circ \frac{\partial \mathbf{r}}{\partial b} \times \frac{\partial \mathbf{r}}{\partial c};$$

or in Cartesian coordinates

$$\rho_0 = \rho \begin{vmatrix} \dfrac{\partial x}{\partial a} & \dfrac{\partial y}{\partial a} & \dfrac{\partial z}{\partial a} \\[2mm] \dfrac{\partial x}{\partial b} & \dfrac{\partial y}{\partial b} & \dfrac{\partial z}{\partial b} \\[2mm] \dfrac{\partial x}{\partial c} & \dfrac{\partial y}{\partial c} & \dfrac{\partial z}{\partial c} \end{vmatrix}.$$

PART III
THE DYNAMICS OF NON-VISCOUS FLUIDS

CHAPTER IX

THE EULERIAN EQUATION AND ITS INTEGRATION ALONG A STREAMLINE

55. General Remarks on the Action of Fluid Viscosity.—The fundamental equation of the dynamics of a particle is:

$$\text{Force} = \text{mass} \times \text{acceleration}.$$

This equation must be satisfied for every element of the fluid. The forces affecting a fluid particle are:

1. The weight per unit volume: $\rho g = \gamma$.
2. The pressure drop per unit volume: $-\text{grad } p$.
3. The friction or viscosity force.

The action of viscosity in fluids is of a rather complicated nature. It is derivable from a symmetric tensor in much the same way as the state of stress of an elastic body, except that in the case of the viscous fluids the stresses are proportional to the velocities, whereas in elasticity they are proportional to the displacements.

Several remarks of a general nature may be made about the action of fluid viscosity: The motion of a fluid is governed essentially by inertia and friction. In most cases we can ignore the effect of gravity, which in compressible homogeneous fluids makes itself noticeable only by a change of density. Experience shows that when we are dealing with large quantities of liquids or gases, the action of friction in the interior of the fluid is much less important than that of inertia. The differences of pressure that occur here are in almost every case due to the forces of inertia. This state of affairs is entirely reversed in dealing with small masses of liquid in which generally the accelerations are very small; here the differences of pressure are balanced almost entirely by frictional forces and the effect of inertia can be neglected. Examples of this are the motion of small drops in clouds, flow through capillary tubes, or the phenomena in lubricated bearings.

It is on account of the mathematical difficulties involved that the general problem of fluid motion was not attacked until a

few years ago, with the exception of a few simple cases. Only by making quite radical assumptions could the equations be so simplified that solutions were obtainable. On one hand, the differential equations simplify so considerably that they can be integrated in many cases if the viscous forces are neglected entirely. This is the so-called "classical hydrodynamics," and the frictionless fluid dealt with is known as the "ideal fluid." This domain has been investigated so thoroughly—particularly by the mathematician—that we may consider it closed. On the other hand, the simplification obtained by completely neglecting inertia also allows of a mathematical treatment. We can therefore develop a scheme of three domains:

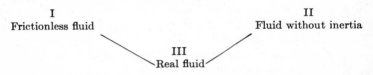

I		II
Frictionless fluid		Fluid without inertia
	III	
	Real fluid	

Forces due to pressure→Forces due to pressure,←Forces due to pressure
  and inertia only   inertia, and friction   and friction only

Approximately twenty-five years ago, advances into domain III from II were made by taking into account the first term of a series development of the inertia forces in the viscous fluid; and also attempts were made to break into III from I by considering a small amount of viscosity in the ideal fluid. The advance from the realm of viscous fluids is associated with the name of Oseen,[1] and the advance from the point of view of the ideal fluid with that of Prandtl,[2] who in 1904 in a paper before the Mathematical Congress at Heidelberg showed how to treat fluids of small viscosity. In this way an impulse was given to the field of hydrodynamics which has shown itself to be very fruitful and of which the consequences are still in a stage of development.

Up to now it has been found impossible to make a mathematical approach to those types of motion in which the frictional and inertia forces are of the same order of magnitude. In a few cases however it has been possible to obtain solutions of the differential equations for viscous fluids, for example in the

[1] OSEEN, C. W., On the Theory of Fluid Resistance (German), *Nova Acta R. Soc. Sci. Uppsala*, ser. 4, vol. 4, 1914.

[2] PRANDTL, L., On Fluid Motions with very Small Viscosity (German), *Proc. 3d Intern. Math. Cong., Heidelberg*, 1904, Leipzig, 1905.

case of a definite simple motion around an infinite plate or for the flow in channels of certain forms. A question presents itself concerning the difference between the actual motion of a fluid with small viscosity (*e.g.*, water) and the theoretical motion of the ideal frictionless fluid of hydrodynamics. It can be stated first that the action of viscosity in liquids of small viscosity is generally confined to a very thin layer next to the boundaries of the rigid body. In this layer—the so-called "Prandtl boundary layer"—there is a steep increase in velocity, to which the frictional force is proportional (see Chap. XV). Although ideal frictionless liquids slide over boundary surfaces, the particles of a fluid of even very small viscosity adhere to the body so that their velocity with respect to the body is zero. When we are dealing with a liquid of small internal friction this change of velocity takes place in a very thin layer, since considerable velocities—such as would exist in the frictionless fluid—appear at a very short distance from the body. The large velocity gradient thus set up is connected with the existence of viscous forces in the boundary layer which are of the same order as the pressure forces.

As long as this thin layer, in which the action of viscosity takes place, remains in contact with the body, the actual streamline pattern does not differ greatly from that derived theoretically on the basis of the ideal frictionless liquid. When however, as occurs most frequently, the streaming motion breaks away from the body, the general nature of the streamline pattern is immediately altered. Part of the boundary layer separates out as a vortex which gives the motion an entirely new character. In this case, the assumption of an ideal frictionless fluid cannot be carried to any experimentally valuable conclusion.

Let us consider the motion about a slender airship or airfoil as an example. Here also the frictional forces are at work in the boundary layer because the velocity is braked off to zero on the body, but since the layer remains clinging to the body and the streaming motion does not "tear away," it does not introduce any essential difference into the form of the streamlines. In the case of motion round a sphere or a flat plate placed at right angles to the stream, however, the boundary layer does not remain in contact with the body but detaches itself in definite places and then proceeds and diffuses as vortices. It is therefore of fundamental importance to find the conditions

under which a boundary layer can, or cannot, detach itself from the body.

In all cases where the boundary layer remains in contact with the body, a useful approximation to the actual flow can be calculated by considering the fluid as frictionless. Then, by decreasing the viscosity more and more, the boundary layer is made thinner and thinner till finally in the limit it disappears at zero viscosity and the streamline pattern becomes that of the ideal frictionless fluid.

In the other case, if we allow the boundary layer sufficient time to develop, separate, and form vortices—we do not get the previous picture of frictionless motion at the limit of zero viscosity.

Since in many cases useful information is obtained from an analysis in which the viscosity is entirely neglected, we shall develop this theory in the next few chapters. The results cannot, however, be accepted unconditionally; they have to be completed by a supplementary investigation to find out whether conditions for the separation of the boundary layer exist or not, *i.e.*, to find out whether the motion can be considered as a good approximation or whether the hypothesis of a frictionless fluid introduces essential differences. We now close our general remarks about the action of viscosity and proceed to the development of the fundamental differential equation of hydrodynamics.

56. Euler's Equation.—The mass of a volume ΔV of density ρ is $\rho \Delta V$. In order to obtain the acceleration we must take the substantial differential coefficient with respect to time since we are dealing with the acceleration of a fluid particle. As we are confining our discussion in this section to frictionless fluids, the forces, according to the statements at the beginning of Art. 55, are the body force $\gamma \Delta V = \rho \mathbf{g} \Delta V$ and the pressure force grad $p \, \Delta V$.

Applying Newton's law to a particle, we find

$$\rho \Delta V \frac{D\mathbf{w}}{dt} = \mathbf{g}\rho \Delta V - \text{grad } p \, \Delta V;$$

or by using Eq. (2) of page 96, and on division by $\rho \Delta V$,

$$\frac{D\mathbf{w}}{dt} = \frac{\partial \mathbf{w}}{\partial t} + \mathbf{w} \circ \text{grad } \mathbf{w} = \mathbf{g} - \frac{1}{\rho} \text{grad } p. \tag{1}$$

This equation of classical hydrodynamics bears the name of "Euler's equation" since Euler first derived it and made it the

foundation upon which he built all his work on hydrodynamics. In coordinate notation we find, on referring to Eq. (2a), page **97**, that the equation becomes

$$\left.\begin{aligned}
\frac{\partial u}{\partial t} + u\frac{\partial u}{\partial x} + v\frac{\partial u}{\partial y} + w\frac{\partial u}{\partial z} &= g_x - \frac{1}{\rho}\frac{\partial p}{\partial x}\\
\frac{\partial v}{\partial t} + u\frac{\partial v}{\partial x} + v\frac{\partial v}{\partial y} + w\frac{\partial v}{\partial z} &= g_y - \frac{1}{\rho}\frac{\partial p}{\partial y}\\
\frac{\partial w}{\partial t} + u\frac{\partial w}{\partial x} + v\frac{\partial w}{\partial y} + w\frac{\partial w}{\partial z} &= g_z - \frac{1}{\rho}\frac{\partial p}{\partial z}
\end{aligned}\right\}. \tag{1a}$$

In the Eulerian equation we therefore have three equations for the five unknowns u, v, w, p, and ρ. Another condition is furnished by the equation of continuity

$$\frac{\partial \rho}{\partial t} + \mathbf{w}\circ \text{grad } \rho + \rho \text{ div } \mathbf{w} = 0.$$

We still lack one equation. In the equation of continuity we made use of an assertion about the conservation of matter, and we shall find the fifth equation by expressing the conservation of energy. This is done by the first law of thermodynamics: The increase in internal energy dU of a system is equal to the sum of the added heat dQ and the work done on the system $-pdV$, so that, if E is the mechanical equivalent of heat,

$$dU = dQ + E(-pdV + \text{work of friction}).$$

Since we are neglecting friction and heat conduction, this simplifies to

$$dU = -EpdV.$$

According to the laws of thermodynamics this condition describes an adiabatic change of state which is satisfied by some relation of the form

$$\rho = f(p).$$

If we assume that our fluid is an ideal ("permanent") gas, then the equation for the adiabatic change of state is

$$pv^k = \text{const.}$$

or since $v = 1/\rho$,

$$p = \rho^k \text{ const.}$$

where

$$k = \frac{c_p}{c_v} \qquad \text{(for air, } k = 1.405\text{)}.$$

When we are dealing with an incompressible fluid, then instead of the equation $\rho = f(p)$, we have $\rho =$ constant.

We therefore see that the motion of a non-viscous fluid is completely determined by the Eulerian equation, the equation of continuity, and a statement about the density derived from the first law of thermodynamics.

The relations become exceedingly complicated in the case of heterogeneous fluids, as for example in the problem of oscillations between layers of different salt solutions. Here another variable—the amount of concentration—enters, so that the density of the various particles is not equal, $\rho = f(\rho_0, a, b, c)$. Such a fluid motion can be treated best by taking the hydrodynamical equations in the Lagrangian form. This form of the equations of motion is obtained as follows: Since the substantial differential of velocity in this system is $(\partial^2 \mathbf{r}/\partial t^2)_s$, we get, on dividing by $\rho \Delta V$,

$$\left(\frac{\partial^2 \mathbf{r}}{\partial t^2}\right)_s = \mathbf{g} - \frac{1}{\rho} \operatorname{grad}_r p,$$

in which the two terms on the right-hand side have still to be transformed to **s**-space. Now

$$\operatorname{grad}_s p = \operatorname{grad}_r p \circ \boldsymbol{\nabla}_s \mathbf{r},$$

is a three-dimensional generalization of

$$\frac{dp}{ds} = \frac{dp}{dx} \frac{dx}{ds},$$

and, if we assume the existence of a force function for **g**,

$$\mathbf{g} = \operatorname{grad}_r U,$$

then

$$\operatorname{grad}_s U = \operatorname{grad}_r U \circ \boldsymbol{\nabla}_s \mathbf{r}.$$

If we insert both these values in the above equation, after having multiplied them scalarly by $\boldsymbol{\nabla}_s \mathbf{r}$, we obtain the fundamental equation of the hydrodynamics of a non-viscous fluid in the Lagrangian form:

$$\boldsymbol{\nabla} \mathbf{r} \circ \frac{\partial^2 \mathbf{r}}{\partial t^2} = \operatorname{grad} U - \frac{1}{\rho} \operatorname{grad} p,$$

where all derivatives are taken in **s**-space.

57. Integration of Euler's Equation along a Streamline.—An integral of great practical importance is the line integral along a streamline **C** of the Euler equation for steady motion. If we

postulate the existence of a force function for the body force **g**, so that **g** = grad U, we have for this integral

$$\int^{\mathbf{c}} \frac{\partial \mathbf{w}}{\partial t} \circ d\mathbf{r} + \int^{\mathbf{c}} (\mathbf{w} \circ \nabla \mathbf{w}) \circ d\mathbf{r} = \int^{\mathbf{c}} \text{grad } U \circ d\mathbf{r} -$$

$$\int^{\mathbf{c}} \frac{\text{grad } p}{\rho} \circ d\mathbf{r} + \text{const.}$$

The integrals on the right-hand side are:

$$\int^{\mathbf{c}} \text{grad } U \circ d\mathbf{r} = \int^{\mathbf{c}} dU = U + \text{const.}$$

$$\int^{\mathbf{c}} \frac{\text{grad } p}{\rho} \circ d\mathbf{r} = \int^{\mathbf{c}} \frac{dp}{\rho} = P(p) + \text{const.}$$

Since we are integrating along a streamline, **w** is parallel to $d\mathbf{r}$, and

$$(\mathbf{w} \circ \nabla \mathbf{w}) \circ d\mathbf{r} = d\mathbf{r} \circ (\mathbf{w} \circ \nabla \mathbf{w}),$$

so that

$$\mathbf{w} \circ (d\mathbf{r} \circ \nabla \mathbf{w}) = \mathbf{w} \circ d\mathbf{w} = d\frac{\mathbf{w}^2}{2}.$$

On account of the importance of this step we shall develop this transformation in Cartesian coordinates also:

$$\mathbf{w} \circ \nabla \mathbf{w} \circ d\mathbf{r}$$

$$= (\mathbf{i}u + \mathbf{j}v + \mathbf{k}w) \circ \left(\mathbf{i}\frac{\partial \mathbf{w}}{\partial x} + \mathbf{j}\frac{\partial \mathbf{w}}{\partial y} + \mathbf{k}\frac{\partial \mathbf{w}}{\partial z} \right) \circ (\mathbf{i}dx + \mathbf{j}dy + \mathbf{k}dz)$$

$$= \left(u\frac{\partial \mathbf{w}}{\partial x} + v\frac{\partial \mathbf{w}}{\partial y} + w\frac{\partial \mathbf{w}}{\partial z} \right) \circ (\mathbf{i}dx + \mathbf{j}dy + \mathbf{k}dz)$$

$$= \left(u\frac{\partial u}{\partial x} + v\frac{\partial u}{\partial y} + w\frac{\partial u}{\partial z} \right) dx + \left(u\frac{\partial v}{\partial x} + v\frac{\partial v}{\partial y} + w\frac{\partial v}{\partial z} \right) dy$$

$$+ \left(u\frac{\partial w}{\partial x} + v\frac{\partial w}{\partial y} + w\frac{\partial w}{\partial z} \right) dz.$$

Since we are integrating along a streamline, *i.e.*, since

$$dx : dy : dz = u : v : w \text{ or } vdx = udy, \qquad wdx = udz, \qquad wdy = vdz,$$

the above expression becomes:

$$\left(u\frac{\partial u}{\partial x}dx + u\frac{\partial u}{\partial y}dy + u\frac{\partial u}{\partial z}dz \right) + \left(v\frac{\partial v}{\partial x}dx + v\frac{\partial v}{\partial y}dy + v\frac{\partial v}{\partial z}dz \right)$$

$$+ \left(w\frac{\partial w}{\partial x}dx + w\frac{\partial w}{\partial y}dy + w\frac{\partial w}{\partial z}dz \right) = udu + vdv + wdw$$

$$= d\left(\frac{u^2 + v^2 + w^2}{2} \right) = d\frac{\mathbf{w}^2}{2}.$$

This example shows the inconvenience of returning to Cartesians, as compared with the direct vector calculation.

The line integral of the Euler equation along a streamline thus is

$$\int^C \frac{\partial \mathbf{w}}{\partial t} \circ d\mathbf{r} + \frac{\mathbf{w}^2}{2} + P - U = \text{const.} \qquad (2)$$

This equation is generally true for non-steady motion but then it must be noted that it holds only at a definite instant of time since at the next instant other fluid particles form the streamline. But even in the case where the streamlines retain their shape but the velocities alter as time goes on, the constant in (2) is variable, *i.e.*, a function of the time. Physically this means that the pressure in the space in which the motion takes place can still be arbitrarily influenced by extraneous causes.

58. Bernoulli's Equation.—For steady motion, *i.e.*, when $\partial \mathbf{w}/\partial t = 0$, the above equation simplifies to

$$\frac{\mathbf{w}^2}{2} + P - U = \text{const.} \qquad (3)$$

This equation is of the utmost importance in the hydrodynamics of a non-viscous fluid and was derived by Daniel Bernoulli in his "Hydrodynamica" of 1738. It is known as "Bernoulli's equation." (The expression for the equilibrium of a homogeneous mass of gas, $P = U + \text{constant}$, obtained in Eq. [6], Art. 3, is a simplified case of Bernoulli's equation with $\mathbf{w}$ equal to zero.)

Let us consider now the special case of incompressible fluids ($\rho = \text{constant}$),[1] with gravity as the body force

$$U = \text{const.} = -gz.$$

Then the Bernoulli equation becomes

$$\frac{\mathbf{w}^2}{2} + \frac{p}{\rho} + gz = \text{const.} \qquad (3a)$$

or on dividing by g and substituting $\rho g = \gamma$,

$$\frac{\mathbf{w}^2}{2g} + \frac{p}{\gamma} + z = \text{const.} \qquad (3b)$$

This form of the Bernoulli equation, which is one in which a simple geometrical meaning can be attached to each term,

[1] It will be shown in Chap XIII that gases can be considered as incompressible fluids if the velocities are small in comparison with the velocity of sound.

is extensively used by engineers. Since z is a length—in this case the height—each of the terms of the equation must also be of the dimensions of a length. With this in view $\mathbf{w}^2/2g$ can be interpreted as the height from which a particle at rest must fall under the influence of gravity in order to attain the velocity $\mathbf{w}$ and is called the "velocity head"; p/γ is the height reached by a column of fluid under the action of a pressure p against gravity; it is known as the "pressure head." Finally z is the height of the point on the streamline above some definite base surface.

The Bernoulli equation in the form $(3b)$ states that the sum of the velocity head, the pressure head, and z is constant at every point of some definite streamline, if we are dealing with steady motion of an incompressible fluid under the influence of a force function.

This constant, the so-called "Bernoulli constant," may be different for each streamline, and it is only in the case of no rotation—as will be seen in Art. 61—that the constants of all streamlines are identical. Such a rotationless motion exists, for example, when all streamlines come from some large basin of liquid in which the velocities are so small that their squares can be neglected, as in the case of the flow of water from a small opening in the lower portion of a large vessel. In that basin the fluid is practically at rest ($\mathbf{w} = 0$), so that Bernoulli's equation simplifies to $P - U =$ constant, or, in the case of gravity with constant ρ,

$$\frac{p}{\gamma} + z = \text{const.}$$

in which the constant at every point of the still water is the same. Since the Bernoulli number is constant along a definite streamline, and since all the streamlines come from a basin in which the number has the same value everywhere, it follows that the number is constant for all streamlines emerging from this basin. The case in which the Bernoulli numbers differ from streamline to streamline occurs when, for example, two fluid masses coming from different vessels meet.

As a result of the hypothesis that ρ is constant everywhere in the fluid, we can give the Bernoulli equation a particularly simple form. In this case we can split the total pressure p into a gravitational pressure $\bar{p}$ caused by the weight of the fluid and a pressure $p^\star$ due to dynamical action. The pressure $p^\star$

is equal to the pressure in a fluid in motion when gravity is neglected. We are primarily interested in the pressure due to dynamical action, *i.e.*, in the pressure differences produced by motion, since the distribution of pressure in liquids at rest is usually of less importance. We therefore express the Bernoulli equation in such a form that this pressure is separated from the gravitational pressure. The gravitational pressure is given by the equation

$$\bar{p} = \text{const.} - \gamma h.$$

The total pressure at a point in the fluid in motion is therefore

$$p = \bar{p} + p^\star = \text{const.} - \gamma h + p^\star.$$

Substituting this into the Bernoulli equation (3*a*), we get

$$\frac{\mathbf{w}^2}{2} + \frac{p^\star}{\rho} = \text{const.}$$

Hence the larger the velocity the smaller the pressure and *vice versa*.

It is seen that on account of the fact that the density is independent of the height, the weight of each particle in the interior of the fluid is neutralized by the buoyancy it experiences from the neighboring particles. We can therefore calculate the motion of a heavy incompressible fluid without considering gravity. Gravity need only be introduced at the free boundary surface where the total pressure p and not the dynamical pressure $p^\star$ must satisfy certain relations.

This simplification of the Bernoulli equation cannot be applied to gases in case the compressibility must be taken into account, since here the density ρ changes with the time under the influence of both statical and dynamical actions. The gravitational and inertia effects of each element depend upon density so that here the pressure cannot be split up as before.

59. Applications of the Bernoulli Equation.—1. *Motion of a Fluid over a Weir* (Fig. 62).—Since the surface of the liquid is a surface of equal pressure, we have from the Bernoulli equation; (3*b*),

$$\frac{\mathbf{w}^2}{2g} + z = \text{const.}$$

If the shape of the surface is known—for instance from a photograph—the velocity at each point of the surface can be calculated from this formula without any difficulty. The only

additional value required is the general velocity $\mathbf{w}_1$ of approach of the water.

In general, if any statements of a geometrical nature can be obtained about a fluid motion, it is immediately possible to calculate further results by means of the Bernoulli equation.

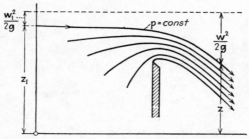

Fig. 62.—Flow over a weir.

2. *Velocity of Efflux from a Small Aperture in an Open Vessel* (Fig. 63).—In this case the geometrical statement consists of the observed fact that the water issues in the form of a jet.

The pressure on the upper surface of the fluid in the vessel and on the surface of the jet is the atmospheric pressure p_0. We assume that the pressure in the interior of the jet is also p_0.

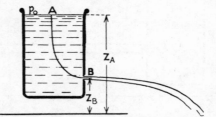

Fig. 63.—Flow from a small opening in an open vessel.

With sufficient accuracy, the velocity $\mathbf{w}_A$ on the free surface A can be considered as being equal to zero; the height of the free surface above some reference level is z_A. The velocity at a point in the jet at height z_B is $\mathbf{w}_B$. Therefore

$$\frac{\mathbf{w}_B{}^2}{2g} + z_B = 0 + z_A,$$

and so

$$\frac{\mathbf{w}_B{}^2}{2g} = z_A - z_B = h,$$

or

$$w_B = \sqrt{2gh}.$$

Since this relation was discovered by Torricelli (a pupil of Galileo) about a hundred years before the Bernoulli equation,

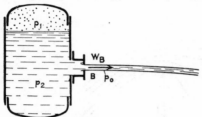

it is known as "Torricelli's theorem."

3. *Velocity of Efflux from a Pressure Tank.*—With the symbols shown in Fig. 64, the Bernoulli equation becomes

$$\frac{\mathbf{w}_B{}^2}{2} + \frac{p_0}{\rho} = 0 + \frac{p_1}{\rho}$$

FIG. 64.—Flow from a gas container. or

$$w_B = \sqrt{2\frac{p_1 - p_0}{\rho}}.$$

The static water head h, which would be equivalent to the pressure difference $p_1 - p_0$, can be found from Eq. (9), page 20:

$$h = \frac{p_1 - p_0}{\gamma} = \frac{p_1 - p_0}{\rho g}.$$

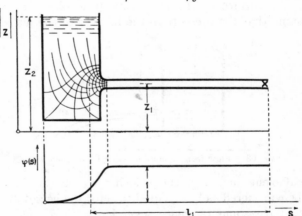

FIG. 65.—Flow through a connecting tube.

This expression substituted into the above equation for w_B gives again

$$w_B = \sqrt{2gh}.$$

4. *Transient Starting Velocity in Drain Pipe* (Fig. 65).—We shall neglect friction although it is of importance when the pipe

is very long. Imagine that the cock at the end of the drain pipe is suddenly opened, how will the velocity of efflux build up in the course of time? Since we are obviously dealing with a non-steady motion, we must use the more general Bernoulli equation of page 114 including the term $\int \partial \mathbf{w}/\partial t \circ d\mathbf{r}$, which since we are dealing with a one-dimensional problem can be written as $\int \frac{\partial w}{\partial t} ds$, where ds is a streamline element. The liquid is assumed to be incompressible.

Let A_1 be the cross section at the end of the pipe, and $A = f(s)$ the section at any point s along the pipe—which need not be uniform. The equation of continuity is:

$$A w = A_1 w_1$$

or

$$w = \frac{A_1}{A} w_1.$$

The quotient A_1/A is a function of s and therefore $w = \varphi(s) w_1$, which gives

$$\frac{\partial w}{\partial t} = \frac{dw_1}{dt} \varphi(s),$$

and

$$\int \frac{\partial w}{\partial t} ds = \frac{dw_1}{dt} \int \varphi(s) ds.$$

We tentatively draw in the cross sections A at right angles to the streamlines, so far back in the vessel that A_1/A is small in comparison with unity. Then $\varphi(s)$ behaves approximately as shown in Fig. 65 for a tube of constant cross section. The value of the integral $\int \varphi(s) ds$, which has the dimensions of a length, is obtained by graphical integration. If the result of this process is l_1, we have

$$\int \frac{\partial w}{\partial t} ds = l_1 \frac{dw_1}{dt}.$$

When the drain pipe has a constant cross section, l_1 is the length of the pipe plus a small increment due to the approach from the tank to the pipe.

The pressure at the end of the tube is that of the atmosphere p_0, and the Bernoulli equation is therefore

$$l_1 \frac{dw_1}{dt} + \frac{w_1^2}{2} + \frac{p_0}{\rho} + g z_1 = C.$$

The equation for a point on the water level in the tank is

$$\frac{p_0}{\rho} + gz_2 = C,$$

and, since the above two constants are equal,

$$l_1\frac{dw_1}{dt} + \frac{w_1^2}{2} = g(z_2 - z_1) = gh.$$

This is the differential equation for w_1. At the commencement of motion, when the velocity w_1 is still small, dw_1/dt is great, but it decreases as w_1 grows. If water is being poured into the tank so as to keep its level constant, the differential equation is of the first degree.

The equation can be written in a somewhat different form by observing that after a sufficient time (theoretically an infinite time), a steady state with velocity $\sqrt{2gh}$ is set up. If we substitute $gh = w_\infty^2/2$ into the above equation, we get

$$\frac{dw_1}{dt} = \frac{w_\infty^2 - w_1^2}{2l_1}$$

or

$$\frac{dw_1}{w_\infty^2 - w_1^2} = \frac{dt}{2l_1},$$

which gives

$$w_1 = w_\infty \tanh \frac{w_\infty t}{2l_1}.$$

The velocity at the end of the drain pipe increases like the hyperbolic tangent, *i.e.*, at first almost linearly, then gradually slower, until finally an asymptotic value is approached.

When there is no flow into the vessel, its level sinks, and if A_0 be the section of the tank at the water level, and A_1 the cross section of the pipe, we have

$$A_0\frac{dh}{dt} = -A_1w_1.$$

If this relation be inserted into the one given above we obtain a differential equation of the second degree which will not be discussed here.

5. *Jet Impinging at Right Angles on a Flat Plate.*—From symmetry the middle stream filament must hit the plate at right angles with zero velocity. Using the notation of Fig. 66, the Bernoulli equations for two points on the middle streamline are:

$$\frac{w^2}{2} + \frac{p_0}{\rho} = C,$$

$$0 + \frac{p_1}{\rho} = C,$$

and, since these constants are identical, we have

$$\frac{w^2}{2} + \frac{p_0}{\rho} = \frac{p_1}{\rho},$$

$$p_1 = p_0 + \frac{\rho w^2}{2},$$

or

$$p_1 = p_0 + \frac{\gamma w^2}{2g}.$$

The point on the plate at which the middle filament comes to rest is called the "stagnation point." The increase in pressure caused by the loss in velocity is $\gamma w^2/2g$ and is called the "stagnation or dynamical pressure." The fluid pressure p_0 is usually called the "static pressure" in the technical literature. It is the force per unit area with which two neighboring fluid particles are pressed together and would be shown by a pressure gauge carried along by the stream. The term "total pressure" is necessary to denote the sum of the static and dynamic pressures.

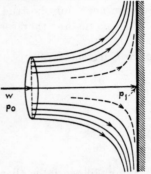

Fig. 66.—Horizontal jet (axisymmetric) at right angle to a plate.

It is to be noted that the last equation is in general only true for homogeneous incompressible fluids ($\rho = $ constant), for which we were able, as on page 116, to eliminate the pressure due to the weight of the fluid itself. If in this case we split the pressure p obtained from the Bernoulli equation

$$p + \frac{\rho}{2}w^2 + \gamma h = \text{const.}$$

into the gravitational (hydrostatic) pressure $\bar{p}$—given by the hydrostatic distribution of pressure $\bar{p} + \gamma h = $ constant—and the pressure $p^\star$ associated with the motion of the fluid and the dynamic action, we realize that the term static pressure for the pressure which is connected with the motion is not very fortunate. It would have been preferable to call $p^\star$ something like "motion pressure."

CHAPTER X

POTENTIAL MOTION

60. Simplification of Euler's Equation, and Integration on Assuming a Velocity Potential.—We shall now form an important and rather general integral of Euler's differential equation. Euler's equation can immediately be written in an integrable form if we assume that the motion is everywhere free from rotation. It should be noted that this assumption of irrotational motion need only be true for some instant of time.

At this instant therefore

$$\text{rot } \mathbf{w} = 0$$

or in Cartesians,

$$\frac{\partial w}{\partial y} - \frac{\partial v}{\partial z} = 0,$$

$$\frac{\partial u}{\partial z} - \frac{\partial w}{\partial x} = 0,$$

$$\frac{\partial v}{\partial x} - \frac{\partial u}{\partial y} = 0.$$

We shall see later that the hypothesis of irrotational motion is in no way such a sweeping limitation of possible motions as appears at first glance; it is in fact implied in many cases of motion of non-viscous fluids.

As was seen in Art. 47, a velocity field $\mathbf{w}$ in which the rotation is everywhere zero can be expressed as the gradient of a scalar Φ.

$$\mathbf{w} = \text{grad } \Phi.$$

This means that in this case there is always a scalar function $\Phi(x, y, z)$, so that

$$u = \frac{\partial \Phi}{\partial x}, \qquad v = \frac{\partial \Phi}{\partial y}, \qquad w = \frac{\partial \Phi}{\partial z}.$$

Φ is called the potential function of the velocity field, and the motion of the liquid is called a potential motion. We shall see later that potential motion is of fundamental importance to hydrodynamics.

Mathematically the assumption of a potential function introduces important simplifications in that we have now to find only one function $\Phi(x, y, z)$ instead of having to deal with three functions: the velocity components $u(x, y, z)$, $v(x, y, z)$, and $w(x, y, z)$.

We now ask: What significance has the assumption of no rotation at the time $t = t_1$ in connection with Euler's equation? Since the hypothesis $\mathbf{w} = \text{grad } \Phi$ is made for only a definite instant of time, we cannot, without further consideration, substitute grad Φ for $\mathbf{w}$ in the term $\partial\mathbf{w}/\partial t$ of Euler's equation. The term $\mathbf{w}\circ \text{grad } \mathbf{w}$ can be given a different form, as follows: We start from the well-known vector relation between any three non-coplanar vectors $\mathbf{a}, \mathbf{b}, \mathbf{c}$:

$$\mathbf{a} \times (\mathbf{b} \times \mathbf{c}) = \mathbf{a}\circ\mathbf{c}\ \mathbf{b} - \mathbf{a}\circ\mathbf{b}\ \mathbf{c}.$$

Putting

$$\mathbf{a} = \mathbf{w}, \qquad \mathbf{b} = \boldsymbol{\nabla}, \qquad \text{and} \qquad \mathbf{c} = \mathbf{w},$$

this relation becomes

$$\mathbf{w} \times (\boldsymbol{\nabla} \times \mathbf{w}) = \mathbf{w}\circ\mathbf{w}\ \boldsymbol{\nabla} - \mathbf{w}\circ\boldsymbol{\nabla}\ \mathbf{w}.$$

Or, since

$$\mathbf{w} \times (\boldsymbol{\nabla} \times \mathbf{w}) = \mathbf{w} \times \text{rot } \mathbf{w},$$

we have

$$\mathbf{w}\circ\boldsymbol{\nabla}\ \mathbf{w} = \mathbf{w}\circ\mathbf{w}\ \boldsymbol{\nabla} - \mathbf{w} \times \text{rot } \mathbf{w}.$$

But

$$\mathbf{w}\circ\mathbf{w}\ \boldsymbol{\nabla} = \boldsymbol{\nabla}\ \mathbf{w}\circ\mathbf{w} = \text{grad } \frac{\mathbf{w}^2}{2},$$

and, at the instant under consideration, rot $\mathbf{w} = 0$; hence

$$\mathbf{w}\circ\boldsymbol{\nabla}\ \mathbf{w} = \mathbf{w}\circ \text{grad } \mathbf{w} = \text{grad } \frac{\mathbf{w}^2}{2}.$$

Naturally we obtain the same results if we use co-ordinates:

$$\boldsymbol{\nabla}\mathbf{w} = \left.\begin{matrix} \mathbf{ii}\dfrac{\partial u}{\partial x} + \mathbf{ij}\dfrac{\partial v}{\partial x} + \mathbf{ik}\dfrac{\partial w}{\partial x} \\[2mm] + \mathbf{ji}\dfrac{\partial u}{\partial y} + \mathbf{jj}\dfrac{\partial v}{\partial y} + \mathbf{jk}\dfrac{\partial w}{\partial y} \\[2mm] + \mathbf{ki}\dfrac{\partial u}{\partial z} + \mathbf{kj}\dfrac{\partial v}{\partial z} + \mathbf{kk}\dfrac{\partial w}{\partial z} \end{matrix}\right\} = \begin{Bmatrix} \dfrac{\partial u}{\partial x} & \dfrac{\partial v}{\partial x} & \dfrac{\partial w}{\partial x} \\[2mm] \dfrac{\partial u}{\partial y} & \dfrac{\partial v}{\partial y} & \dfrac{\partial w}{\partial y} \\[2mm] \dfrac{\partial u}{\partial z} & \dfrac{\partial v}{\partial z} & \dfrac{\partial w}{\partial z} \end{Bmatrix};$$

but, since

$$u = \frac{\partial \Phi}{\partial x}, \qquad v = \frac{\partial \Phi}{\partial y}, \qquad w = \frac{\partial \Phi}{\partial z},$$

we have

$$\frac{\partial u}{\partial y} = \frac{\partial v}{\partial x}\left(= \frac{\partial^2 \Phi}{\partial x \partial y}\right),$$

$$\frac{\partial u}{\partial z} = \frac{\partial w}{\partial x}\left(= \frac{\partial^2 \Phi}{\partial x \partial z}\right),$$

$$\frac{\partial v}{\partial z} = \frac{\partial w}{\partial y}\left(= \frac{\partial^2 \Phi}{\partial y \partial z}\right).$$

We therefore see that we are dealing with a symmetric tensor, $\nabla \mathbf{w}$ which, as we have seen in Art. 44, is the expression for a deformation velocity without rotation. If we now form the x-component of the vector $\mathbf{w} \circ \nabla \mathbf{w}$, we get

$$(\mathbf{w} \circ \nabla \mathbf{w})_x = u\frac{\partial u}{\partial x} + v\frac{\partial u}{\partial y} + w\frac{\partial u}{\partial z}$$

or, since

$$\frac{\partial u}{\partial y} = \frac{\partial v}{\partial x}$$

and

$$\frac{\partial u}{\partial z} = \frac{\partial w}{\partial x},$$

this becomes

$$(\mathbf{w} \circ \nabla \mathbf{w})_x = u\frac{\partial u}{\partial x} + v\frac{\partial v}{\partial x} + w\frac{\partial w}{\partial x} = \frac{\partial}{\partial x}\frac{u^2 + v^2 + w^2}{2}.$$

If we form the corresponding expressions for the y- and z-components, we see that

$$\mathbf{w} \circ \nabla \mathbf{w} = \left(\mathbf{i}\frac{\partial}{\partial x} + \mathbf{j}\frac{\partial}{\partial y} + \mathbf{k}\frac{\partial}{\partial z}\right)\left(\frac{u^2 + v^2 + w^2}{2}\right)$$

$$= \operatorname{grad}\left(\frac{u^2 + v^2 + w^2}{2}\right) = \operatorname{grad}\frac{\mathbf{w}^2}{2}.$$

If now we assume that the body force $\mathbf{g}$ possesses a force function U,

$$\mathbf{g} = \operatorname{grad} U,$$

and further, that the fluid is homogeneous, *i.e.*, that the density ρ is only a function of the pressure p (the fluid may be compressible),

$$\int \frac{dp}{\rho} = P(p)$$

or

$$\frac{1}{\rho}\operatorname{grad} p = \operatorname{grad} P,$$

then Euler's equation becomes

$$\frac{\partial \mathbf{w}}{\partial t} = -\text{grad}\left(\frac{\mathbf{w}^2}{2} + P - U\right).$$

We therefore realize that when the velocity field is irrotational at a definite instant of time t_1, the Euler equation (with the above assumptions about force field and density), shows that at this instant the acceleration field $\partial \mathbf{w}/\partial t$ can be represented as the gradient of the scalar $-(\mathbf{w}^2/2 + P - U)$ and is therefore irrotational at all times.

If we expand the velocity at some definite point as a Taylor series with the time as variable,

$$\mathbf{w}_2 = \mathbf{w}_1 + (t_2 - t_1)\left(\frac{\partial \mathbf{w}}{\partial t}\right)_{t_1} + \frac{(t_2 - t_1)^2}{2!}\left(\frac{\partial^2 \mathbf{w}}{\partial t^2}\right)_{t_1} + \cdots,$$

we have, for a sufficiently small interval $t_2 - t_1$,

$$\mathbf{w}_2 = \mathbf{w}_1 + (t_2 - t_1)\left(\frac{\partial \mathbf{w}}{\partial t}\right)_{t_1}.$$

Since, as was seen above, the acceleration field $\partial \mathbf{w}/\partial t$ is irrotational at the instant t_1, $\mathbf{w}_2$ is also irrotational at t_2. An immediate application of this statement leads to the theorem: If at some definite instant of time the distribution of velocity in a homogeneous non-viscous fluid is irrotational, it will always remain irrotational under the action of a rotation-free force system. This theorem is due to Lagrange. The theorem can be interpreted as stating that it is impossible to impart rotation to an irrotational fluid through the action of a rotation-free force. Future considerations will show that this theorem is also valid for the fluid in a finite region when the motion in this region is irrotational to start with. It is however possible that instabilities at the boundary of this region may produce rotation or instability in the interior of the fluid (see Art. 84 and the concluding remarks of the previous chapter). The limitation imposed on the theorem by postulating rotation-free force systems is not very important since rotational fields of force hardly ever occur. One of the exceptional cases of rotational fields is that of the magnetic forces produced under the influence of electric currents passing through the liquid.

The above stated considerations are not valid for those types of motion in which the value of the density is affected by other factors than the pressure only, like local heating for example.

We cannot, therefore, use the Euler equation in its above form if we wish to study the motion of a gas which is being heated in certain places and is brought into motion through the resulting changes in density.

We have thus shown that if the velocity field of a fluid possesses a potential at some instant of time, it will always possess one when certain other assumptions are satisfied. On account of this we are now justified in replacing the velocity in the first term of Euler's equation by grad Φ. Since space integration and time integration are independent, we can write

$$\frac{\partial \mathbf{w}}{\partial t} = \frac{\partial}{\partial t} \operatorname{grad} \Phi = \operatorname{grad} \frac{\partial \Phi}{\partial t},$$

so that finally we can write the Euler equation in the form

$$\operatorname{grad} \left(\frac{\partial \Phi}{\partial t} + \frac{\mathbf{w}^2}{2} + P - U \right) = 0 \qquad (1)$$

and, on integration,

$$\frac{\partial \Phi}{\partial t} + \frac{\mathbf{w}^2}{2} + P - U = f(t), \qquad (2)$$

where $f(t)$ is an arbitrary function of time.

At any instant therefore the expression

$$\frac{\partial \Phi}{\partial t} + \frac{\mathbf{w}^2}{2} + P + U$$

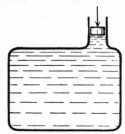

is constant for all particles of the fluid. Since, however, the equation only says that the space differential of the expression in brackets is zero, it is obvious that the constant alters with time, that is, it will in general be a function of time. Physically this means that the pressure in the space in which the motion occurs can still be changed at will by outside influences, as, for example,

Fig. 67.—External forces (piston action) may produce changes of pressure in the fluid motion.

by a piston (see Fig. 67). If, however, we are dealing with an infinite fluid region in which it is impossible to introduce external influences to affect the pressure, the constant is independent of time.

61. Connection between the Integral of Euler's Equation for Potential Motion and the Corresponding Integral along a Streamline.—In Art. 57 we derived a special integral of Euler's equation which was valid only along a streamline, but in Eq. (2), above,

we found a very general integral of the same equation which is true for all points of the liquid, when certain conditions are fulfilled.

The most important of these conditions is that the motion must be irrotational, if only for an instant. To this class belong all motions which are developed from a state of rest, in which state the motion is certainly irrotational (Φ = constant). We can make the general statement that all motions of a homogeneous non-viscous fluid generated from a state of rest are irrotational if the force field possesses a force function (which is nearly always true).

Now we recognize a relation with the Bernoulli equation which is an integral of the Euler equation along a streamline that was also found to be valid for all points of the liquid originating from a basin of such dimensions that the velocity there was practically zero (the Bernoulli number was then identical for all streamlines). In fact it is seen that for the steady state both equations are the same. This means that for steady motion of a non-viscous homogeneous fluid originally at rest the moving fluid comes from a region in which statical conditions prevail, *i.e.*,

$$\frac{\mathbf{w}^2}{2} + P - U = f(t).$$

We therefore know that a motion in which the Bernoulli number is the same for all streamlines is a potential motion and therefore irrotational.

For non-steady motions we cannot integrate Euler's equation along a streamline; the term $\int \frac{\partial \mathbf{w}}{\partial t} \circ d\mathbf{r}$ prevents us from doing that. However, in case the liquid comes from a region at rest, the Bernoulli constant is the same over the whole liquid, and, since we have realized that this means a potential motion, we can now calculate the integral $\int \frac{\partial \mathbf{w}}{\partial t} \circ d\mathbf{r}$ as follows:

$$\int \frac{\partial \mathbf{w}}{\partial t} \circ d\mathbf{r} = \frac{\partial}{\partial t} \int \operatorname{grad} \Phi \circ d\mathbf{r} = \frac{\partial \Phi}{\partial t}.$$

Thus the line integral becomes

$$\frac{\partial \Phi}{\partial t} + \frac{\mathbf{w}^2}{2} + P - U = f(t),$$

which we shall call the general Bernoulli equation.

The important theorem of Lagrange, stating that rotation cannot be produced in an irrotational motion of a homogeneous non-viscous fluid by the action of an irrotational force field (*i.e.*, a field possessing a potential) does not seem to agree with experiment.   It is true that there are no absolutely non-viscous fluids, but in many cases the viscosity is so small that one wonders how such small viscosities as of water or air can change the type of motion so completely.   The answer to this is that the theorem of Lagrange is valid only in regions where the viscous forces can be neglected, and this is the case only in the interior of a liquid (outside the boundary layers) as was explained in Art. 55.   In the thin layer along the boundary surface of the liquid the viscous force is quite considerable even when the viscosity itself is very small.   Here, where the important assumption of absence of viscosity is not even approximately satisfied, Lagrange's theorem is inapplicable.   As was explained in Art. 55, this boundary layer can leave the surface under certain circumstances.   It can then penetrate into the interior of the fluid mass and there completely alter the type of flow in spite of the fact that the fluid is nearly frictionless.   In order to be able to apply the theorem of Lagrange it is absolutely essential that the assumptions of absence of viscosity, absence of rotation, and homogeneity of fluid are satisfied.

62. Equations Defining Potential Functions and Pressure Functions.—In the general Bernoulli equation, (2), and the continuity equation we have two equations for Φ and P from which we can determine Φ.   They are:

$$\frac{\partial \rho}{\partial t} + \rho \operatorname{div} \mathbf{w} + \mathbf{w} \circ \operatorname{grad} \rho = 0 \text{ (general equation of continuity)}$$

and

$$\frac{\partial \Phi}{\partial t} + \frac{\mathbf{w}^2}{2} + P - U = f(t) \text{ (general Bernoulli equation)},$$

in which the density is a function of the pressure only.

Dividing the continuity equation by P and remembering that

$$\operatorname{div} \mathbf{w} = \boldsymbol{\nabla} \circ \mathbf{w} = \boldsymbol{\nabla} \circ \operatorname{grad} \Phi = \boldsymbol{\nabla} \circ \boldsymbol{\nabla} \Phi = \Delta\Phi,$$

we obtain

$$\frac{1}{\rho}\frac{\partial \rho}{\partial t} + \Delta\Phi + \frac{1}{\rho} \operatorname{grad}\Phi \circ \operatorname{grad}\rho = 0.$$

For the velocity of sound c we have the general formula

$$c = \sqrt{\frac{E}{\rho}},$$

where E is the modulus of elasticity, for which we can write

$$E = \frac{dp}{\frac{-dV}{V}};$$

and, since

$$-\frac{dV}{V} = \frac{d\rho}{\rho},$$

we find

$$\frac{E}{\rho} = \frac{dp}{d\rho} = c^2.$$

Also

$$dP = \frac{dp}{\rho} = \frac{dp}{d\rho}\frac{d\rho}{\rho} = c^2\frac{d\rho}{\rho},$$

and therefore the equation of continuity becomes

$$\frac{1}{c^2}\frac{\partial P}{\partial t} + \Delta\Phi + \frac{\text{grad }\Phi \circ \text{grad }P}{c^2} = 0. \tag{3}$$

With the equation of Bernoulli

$$\frac{\partial\Phi}{\partial t} + \frac{\mathbf{w}^2}{2} + P - U = f(t),$$

we have two equations for Φ and P by which every irrotational motion of a homogeneous non-viscous fluid is uniquely determined, subject to the boundary conditions.

This complicated system of equations can be simplified in the following cases:

1. Incompressible fluids ($\rho = \text{constant}$);
2. When the velocity $\mathbf{w}$ is very small;
3. For steady motion:
   a. In large volumes of fluid but with moderately large velocities,
   b. In small volumes of fluid but with very large velocities;
4. In the case of the one-dimensional problem.

63. The Potential Function for Incompressible Fluids.—The equation of continuity here is

$$\text{div }\mathbf{w} = \Delta\Phi = 0. \tag{4}$$

Since this equation only involves Φ and since Φ determines the whole velocity field, the equation $\Delta\Phi = 0$ contains the kinematics of all motions of this type, subject of course to boundary conditions.

The general Bernoulli equation is only of use for subsequently calculating the pressure (when that pressure is prescribed at one point and when the function $f(t)$ is uniquely defined). It is, moreover, necessary that the velocities are prescribed on all boundaries. If this is not so, as in the case of a free surface where no velocity conditions are given, the motion can *not* be determined from Eq. (4) alone. Likewise one cannot divide the calculations into separate determinations of velocity and pressure in cases where the density is not constant.

The equation $\Delta\Phi = 0$, which is called "Laplace's differential equation," has been thoroughly and completely investigated, since it plays an important part not only in hydrodynamics but also in the theory of electric and magnetic fields, as well as in many other physical phenomena. Its discussion forms the subject matter of potential theory and of classical hydrodynamics. Since many solutions of this equation are known and since everything related to it can be found in the standard books,[1] we shall not go into the equation here. Only one remark should be made.

Until now we have encountered only second-degree differential equations (the Euler and the general Bernoulli equation), but here, in dealing with the potential motion of incompressible fluids, a *linear* equation in Φ enters. This involves an important mathematical simplification in that every linear combination of known solutions is also a solution, and so it is made much more easy to satisfy boundary conditions.

64. The Potential Function When the Velocity w Is Very Small. If **w** is so small that $(\mathbf{w} \circ \operatorname{grad} P)/c^2$ in the equation of continuity and $\mathbf{w}^2/2$ in the Bernoulli equation can be neglected as quantities of the second order, the continuity relation becomes

$$\frac{2}{c^2} \frac{\partial P}{\partial t} + \Delta\Phi = 0.$$

If we assume further that the fluid domain is sufficiently restricted so that changes in density produced by the force field can be

[1] See LAMB, H., "Hydrodynamics," Cambridge University Press, 5th ed., 1924.

neglected (this rules out atmospheric phenomena), the general Bernoulli equation becomes

$$\frac{\partial \Phi}{\partial t} + P = \text{const.}$$

The function $f(t)$ has been immediately put equal to a constant since we are excluding the case where the pressure in the interior of the fluid is affected by external influences, (such as for instance by an externally propelled piston on the boundary of the fluid). Differentiating the second equation and substituting

$$\frac{\partial P}{\partial t} = -\frac{\partial^2 \Phi}{\partial t^2}$$

into the equation of continuity, we obtain

$$c^2 \Delta \Phi = \frac{\partial^2 \Phi}{\partial t^2}. \tag{5}$$

This is the well-known equation for the propagation of sound in a homogeneous medium. We shall not discuss the various methods of integrating this equation since everything related to it can be found in the literature on this subject.[1]

We shall, however, give just one brief example. A solution of the differential equation, (5), is given by

$$\Phi = A e^{i(\alpha x - \beta t)},$$

where the complex form must be so interpreted that both the real and imaginary parts of $A e^{i(\alpha x - \beta t)}$ are solutions of the equation. This can be easily verified by substitution.

The velocity components are

$$u = \frac{\partial \Phi}{\partial x} = i\alpha A e^{i(\alpha x - \beta t)},$$

$$v = \frac{\partial \Phi}{\partial y} = 0,$$

$$w = \frac{\partial \Phi}{\partial z} = 0.$$

Forming

$$\Delta \Phi = -\alpha^2 A e^{i(\alpha x - \beta t)}$$

and

$$\frac{\partial^2 \Phi}{\partial t^2} = -\beta^2 A e^{i(\alpha x - \beta t)}$$

[1] RAYLEIGH, "Theory of Sound," vol. 2, chap. XIII.

enables us to write for the differential equation

$$c^2\alpha^2 = \beta^2;$$

hence, if α (and therefore the wave length $\lambda = 2\pi/\alpha$) is fixed, β is given by

$$\beta = \pm c\alpha.$$

We therefore have the two solutions

$$\Phi_1 = Ae^{i[\alpha(x-ct)]}$$

and

$$\Phi_2 = Ae^{i[\alpha(x+ct)]},$$

which represent two waves advancing with the velocity c to the right and left, respectively. We now find ourselves within the sphere of acoustics; all sound waves irrespective of their frequency have the same velocity of propagation. Thus there is no dispersion as occurs, for example, with light waves in isotropic media or with water-surface waves.

Spherical and cylindrical waves travel in much the same way as the plane waves discussed above. For further details the reader is referred to the textbooks on classical theoretical physics.

In connection with the assumptions made in deriving Eq. (5) it is worth while bearing in mind a point about U, namely that for very large vertical distances the density is dependent on the force field, so that c is no longer constant but a function of the height. In addition, we lay great stress on the fact that (5) is to be considered as an approximation valid only for small velocities which for the high frequencies of acoustics is equivalent to small *amplitudes* of vibration. The equation is not true for waves of great amplitude such as are produced, for instance by explosions, where pressure differences of many atmospheres and velocities far beyond that of sound may occur. We shall discuss this subject briefly under case b in the next article.

65. The Potential Function for Steady Motion.—In the steady-state case, when the time derivatives disappear, the Bernoulli equation is of the form

$$\frac{\mathbf{w}^2}{2} + P - U = \text{const.}$$

From this we calculate the expression for grad P and insert it into the continuity equation for steady motion with the result that

$$\Delta\Phi + \frac{\mathbf{w}\circ \text{grad}\left(U - \dfrac{\mathbf{w}^2}{2}\right)}{c^2} = 0. \tag{6}$$

Two cases arise:

(a) The velocities are not too large, so that w^2/c^2 can be neglected; and the fluid or gas domain is so large that changes in density produced by the field of force (the weight) of the gas are to be taken into account.

(b) The velocities are of the order of that of sound but the fluid domain under consideration is so small that changes of density produced by the weight of the fluid can be neglected.

In case a the above equation simplifies to

$$c^2\Delta\Phi + \mathbf{w}\circ \text{grad}\, U = 0. \tag{7}$$

This is the equation of continuity for a homogeneous free atmosphere, true only if the earth's rotation is of no importance. This rotation, however, cannot be neglected when considering volumes of gas spread out horizontally over large areas. Then it must be taken into account so that the problem is no longer one of potential flow. If we assume a homogeneous gravity field, *i.e.*, one in which we neglect the variation of g with height ($U = \text{const.} -gz$) Eq. (7) becomes

$$c^2\Delta\Phi - g\frac{\partial\Phi}{\partial z} = 0. \tag{7a}$$

This equation has not been discussed much. The reason for this is connected with the objection that can be made against it from the physical point of view, namely, that changes in the state of the atmosphere are generally not adiabatic. This results in the atmosphere being composed of layers of unequal density, which contradicts the important assumption of homogeneity necessary for the application of the general Bernoulli equation. However, if one wishes to investigate the motion of an adiabatic atmosphere over a region of moderate height, an approximation can be made by first assuming incompressibility and deriving a function Φ_1 from the equation $\Delta\Phi_1 = 0$. This function is inserted in the term $g\dfrac{\partial\Phi}{\partial z}$ of (7a) and the resulting equation

$$c^2\Delta\Phi_2 = g\frac{\partial\Phi_1}{\partial z}$$

is used as one for determining Φ_2. Φ_2 is now used to determine some Φ_3, etc., and so we approximate more and more closely to the solution of $(7a)$ if of course the process converges and the difficulties involved in satisfying the boundary conditions can be overcome.

In case b the differential equation for a compressible fluid is

$$\Delta\Phi - \frac{\mathbf{w} \circ \operatorname{grad} \dfrac{\mathbf{w}^2}{2}}{c^2} = 0. \tag{8}$$

We assume that the change of density produced by the weight of the gas can be neglected—an atmospheric height of 100 ft only produces a change of about 0.3 per cent. This differential equation can be applied to the case of a gas issuing from some vessel at high pressure (10 atmospheres, for instance) through a hole or tube. Similar relations occur in the theory of steam turbines, and it is really in connection with the theory of this machine that the differential equation was first investigated.

If we express the term $\mathbf{w} \circ \operatorname{grad} \mathbf{w}^2/2$ in Cartesian coordinates, we have

$$\begin{aligned}
\mathbf{w} \circ \operatorname{grad} \frac{\mathbf{w}^2}{2} &= (\mathbf{i}u + \mathbf{j}v + \mathbf{k}w)\circ\left(\mathbf{i}\frac{\partial\dfrac{\mathbf{w}^2}{2}}{\partial x} + \mathbf{j}\frac{\partial\dfrac{\mathbf{w}^2}{2}}{\partial y} + \mathbf{k}\frac{\partial\dfrac{\mathbf{w}^2}{2}}{\partial z}\right) \\
&= \left(u\frac{\partial}{\partial x} + v\frac{\partial}{\partial y} + w\frac{\partial}{\partial z}\right)\left(\frac{u^2 + v^2 + w^2}{2}\right) \\
&= u^2\frac{\partial u}{\partial x} + uv\left(\frac{\partial u}{\partial y} + \frac{\partial v}{\partial x}\right) + v^2\frac{\partial v}{\partial y} + uw\left(\frac{\partial u}{\partial z} + \frac{\partial w}{\partial x}\right) \\
&\qquad + w^2\frac{\partial w}{\partial z} + vw\left(\frac{\partial v}{\partial z} + \frac{\partial w}{\partial y}\right).
\end{aligned}$$

Since further

$$\Delta\Phi = \frac{\partial u}{\partial x} + \frac{\partial v}{\partial y} + \frac{\partial w}{\partial z},$$

Eq. (8) becomes

$$\begin{aligned}
\frac{\partial u}{\partial x}\left(1 - \frac{u^2}{c^2}\right) + \frac{\partial v}{\partial y}\left(1 - \frac{v^2}{c^2}\right) + \frac{\partial w}{\partial z}\left(1 - \frac{w^2}{c^2}\right) - \left(\frac{\partial u}{\partial y} + \frac{\partial v}{\partial x}\right)\frac{uv}{c^2} \\
- \left(\frac{\partial u}{\partial z} + \frac{\partial w}{\partial x}\right)\frac{uw}{c^2} - \left(\frac{\partial v}{\partial z} + \frac{\partial w}{\partial y}\right)\frac{vw}{c^2} = 0. \quad (8a)
\end{aligned}$$

In connection with this equation it must be said that its inde-
pendence of the Bernoulli equation is only apparent, for the
velocity of sound c is a function of pressure or of density. Combining it with the Bernoulli equation it here becomes a function
of $u^2 + v^2 + w^2$.

A solution of this equation can be obtained by successive
approximations when u, v, w are fairly large but are still smaller
than the velocity of sound. The first approximation Φ_1 is
obtained from the equation for an incompressible fluid $\Delta\Phi_1 = 0$
and from it the quantities u_1, v_1, w_1 are derived. With these
values one forms the equation

$$c^2\Delta\Phi_2 = u_1{}^2\frac{\partial u_1}{\partial x} + u_1 v_1\left(\frac{\partial u_1}{\partial y} + \frac{\partial v_1}{\partial x}\right) + \cdots$$

We now obtain Φ_2 and from it u_2, v_2, w_2. In general the smaller
the ratio of w to the velocity of sound c, the more rapidly does
the process converge. This process does not converge when
velocities equal to, or greater than that of sound occur, as is
the case in the efflux of gases from containers.

It is possible to transform the non-linear equation (8a) into
a linear one. In order to do this, we change the relation between
dependent and independent variable by means of a Legendre
transformation, so that u, v, w are now the independent and
x, y, z the dependent variables. The bracketed expressions
$(1 - u^2/c^2)$, etc., now contain the independent variables and
we obtain a linear differential equation, all of whose coefficients
are functions of the independents. Without going into the
details of this transformation it may be briefly mentioned that
in the course of the transformation, we find an expression

$$\Psi = ux + vy + wz - \Phi$$

in the place of the potential Φ. It is seen that this formula is
symmetrical in Φ and Ψ. Combined with the equations

$$u = \frac{\partial\Phi}{\partial x}, \qquad v = \frac{\partial\Phi}{\partial y}, \qquad w = \frac{\partial\Phi}{\partial z},$$

we obtain the new relations

$$x = \frac{\partial\Psi}{\partial u}, \qquad y = \frac{\partial\Psi}{\partial v}, \qquad z = \frac{\partial\Psi}{\partial w}.$$

If these new variables are inserted in Eq. (8a) we get a linear
partial differential equation for (8a) which is of the elliptic type,
when $w < c$, and of hyperbolic type, when $w > c$.

Physically these two equations represent quite different types of motion. Consider the two-dimensional motion over a flat plate, at some point of which there is a corrugation which disturbs the otherwise rectilinear motion of the liquid (Figs. 68 and 69). If the velocity is smaller than that of sound, the upper streamlines quickly smooth themselves out and the disturbance dies down asymptotically outward (Fig. 68). If on the other hand

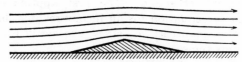

Fig. 68.—Motion over a corrugation in a flat plate when the velocity is less than that of sound. The disturbances die down.

the velocity is larger than that of sound, the corrugations of the wall are propagated into the interior of the fluid at full strength and there is no diminution (Fig. 69). Accordingly we can say that at velocities less than that of sound the fluid disturbances die out, but at velocities greater than that of sound all disturbances at the walls are propagated into the interior of the liquid.

We shall also explain the properties of these two types of motion for the case of a center of disturbance A moving with

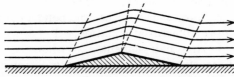

Fig. 69.—Motion over a corrugation in a flat plate when the velocity is greater than that of sound. The disturbances are propagated at full strength into the fluid.

velocity **w** in a fluid extending to infinity in all directions. A momentary disturbance spreads in the form of a spherical wave whose center moves with a velocity **w**. If this velocity is less than c, then obviously after a time t the spherical wave has a position relative to A as shown in Fig. 70. If disturbances continually emanate from A so that they can be considered as momentary disturbances following each other at very short intervals of time, we get the system of waves shown in Fig. 70. The disturbance therefore spreads in all directions, with different velocities in different directions.

If the velocity is greater than c, however, then the spherical wave front of an instantaneous disturbance at a time τ after that disturbance has a position with respect to A as shown in Fig. 71. If there are successive disturbances, the spherical

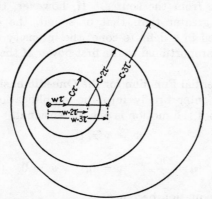

Fig. 70.—The propagation of a disturbance from a point source when its velocity is less than that of sound.

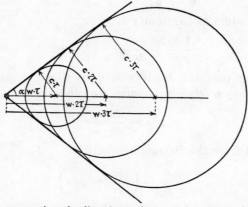

Fig. 71.—The propagation of a disturbance from a point source when its velocity is greater than that of sound.

wave fronts sweep out a cone in space with its vertex at A. There is no disturbance in the space outside the cone. From Fig. 71 we see that the semivertex angle α is determined by the relation

$$\sin \alpha = \frac{c}{w}.$$

Hence a continuous disturbance issuing with the velocity of sound from a point source in a medium which is streaming past with a velocity less than that of sound spreads in all directions, and the intensity of the disturbance is proportional to the square of the distance from the source. If, however, the velocity of the stream is greater than that of sound, the disturbance is only appreciated in a definite cone; the intensity on the surface of that cone is proportional to the first power of the distance from the vertex.

66. The Potential Function for the One-dimensional Problem. We shall now enter briefly into the simplifications introduced by assuming that the motion is one dimensional. We then have

$$\Phi = \Phi(x, t)$$

and

$$u = \frac{\partial \Phi}{\partial x}, \qquad v = 0, \qquad w = 0.$$

The Bernoulli equation becomes

$$\frac{\partial \Phi}{\partial t} + \frac{u^2}{2} + \int \frac{dp}{\rho} = \text{const.},$$

and the equation of continuity is

$$\frac{1}{\rho} \frac{\partial \rho}{\partial t} + \frac{\partial u}{\partial x} + \frac{u}{\rho} \frac{\partial \rho}{\partial x} = 0.$$

If we again postulate homogeneity, by which we mean dependence of density solely on pressure, we can write

$$\frac{d\rho}{\rho} = \frac{d\rho}{dp} \frac{dp}{\rho}.$$

We also get from the Bernoulli equation

$$\frac{dp}{\rho} = -\left(d\frac{\partial \Phi}{\partial t} + u\, du \right),$$

so that the equation of continuity becomes

$$-\left(\frac{\partial^2 \Phi}{\partial t^2} + u\frac{\partial u}{\partial t} \right) \frac{\partial \rho}{\partial p} + \frac{\partial u}{\partial x} - \left(u\frac{\partial^2 \Phi}{\partial x \partial t} + u^2\frac{\partial u}{\partial x} \right) \frac{\partial \rho}{\partial p} = 0$$

or

$$\frac{\partial^2 \Phi}{\partial t^2} + 2\frac{\partial \Phi}{\partial x} \cdot \frac{\partial^2 \Phi}{\partial x \partial t} - \left[\frac{\partial p}{\partial \rho} - \left(\frac{\partial \Phi}{\partial x} \right)^2 \right]\frac{\partial^2 \Phi}{\partial x^2} = 0.$$

By introducing a Legendre transformation with $\partial\Phi/\partial x = u$ as the new independent variable, this equation becomes linear

with respect to t. The further developments of this problem can be obtained on consulting the works of Riemann.[1]

67. Simple Examples of Potential Motion for Incompressible Fluids.—In this article we shall deal with solutions of the equation $\Delta\Phi = 0$. We shall write down some simple expressions for Φ which satisfy the equation $\Delta\Phi = 0$ (this of course should be verified), and we shall then investigate the kind of motion, the form of the streamlines, and the pressure distribution which result from these Φ-functions.

Case 1.

$$\Phi = ax + by + cz,$$

where a, b, c are constants or functions of time. We have

$$u = \frac{\partial\Phi}{\partial x} = a,$$

$$v = \frac{\partial\Phi}{\partial y} = b,$$

$$w = \frac{\partial\Phi}{\partial z} = c,$$

which means that the velocity vector $\mathbf{w} = \mathbf{i}u + \mathbf{j}v + \mathbf{k}w$ is the same at every point of the fluid; we are therefore dealing with a translation of the fluid. If a, b, c, moreover, are independent of the time, we have a steady state of uniform translation. Since such a motion can be transformed to rest by the choice of a suitable moving system of coordinates, the term $\mathbf{w}^2/2$ in the Bernoulli equation

$$\frac{\mathbf{w}^2}{2} + \frac{p}{\rho} - gz = \text{const.}$$

can be removed, and we obtain the fundamental static equation

$$\frac{p}{\rho} - gz = \text{const.}$$

If a, b, c are dependent on time, the general Bernoulli equation is

$$\frac{\partial\Phi}{\partial t} + \frac{\mathbf{w}^2}{2} + \frac{p}{\rho} = f(t),$$

which by considering that

$$\frac{\partial\Phi}{\partial t} = x\frac{da}{dt} + y\frac{db}{dt} + z\frac{dc}{dt},$$

[1] RIEMANN-WEBER, "The Differential Equations of Physics" (German), vol. 2, p. 507, 5th ed., Brunswick, 1912.

can be written as

$$\frac{p}{\rho} = f(t) - \left(x\frac{da}{dt} + y\frac{db}{dt} + z\frac{dc}{dt} \right),$$

where $\mathbf{w}^2/2$, being only a function of t, has been absorbed in $f(t)$. The pressure is seen to increase linearly with the distance, corresponding to the acceleration given to the individual particles.

We now form the pressure gradient grad $p = \mathbf{i}\frac{\partial p}{\partial x} + \mathbf{j}\frac{\partial p}{\partial y} + \mathbf{k}\frac{\partial p}{\partial z}$, with the result:

$$\text{grad } p = \rho\left(\mathbf{i}\frac{da}{dt} + \mathbf{j}\frac{db}{dt} + \mathbf{k}\frac{dc}{dt} \right).$$

If a, b, c are the three components of some vector

$$\mathbf{a} = \mathbf{i}a + \mathbf{j}b + \mathbf{k}c,$$

the last result can be written as

$$\text{grad } p = \rho\frac{d\mathbf{a}}{dt}.$$

In investigating the form of the streamlines, we get

$$dx:dy:dz = a:b:c;$$
$$\frac{dy}{dx} = \frac{b}{a} = \text{const.}$$

or

$$y = \frac{b}{a}x + \text{const.};$$

and again

$$\frac{dz}{dy} = \frac{c}{b} = \text{const.}$$

or

$$z = \frac{c}{b}y + \text{const.},$$

so that the streamlines are straight.

A case of translatory non-steady motion occurs for example when a vessel entirely filled with liquid is moved parallel to itself in some curved path. This is also an example of the case where streamlines and path lines are not identical. Although the streamlines are, as we have seen, straight, the path lines have the form prescribed by the curved path of the vessel.

Case 2.

$$\Phi = \tfrac{1}{2}(ax^2 + by^2 + cz^2).$$

Since $\Delta\Phi = a + b + c$, we get a solution of the equation $\Delta\Phi = 0$ when $a + b + c = 0$. This condition can be satisfied in many ways.

We shall first consider the case (a)

$$b = -a, \qquad c = 0,$$

so that

$$\Phi = \frac{a}{2}(x^2 - y^2).$$

Since this equation is independent of z we are dealing with a two-dimensional motion. By differentiating the potential function we find for the velocity components

$$u = ax, \qquad v = -ay, \qquad w = 0.$$

The form of the streamlines is thus obtained quite simply,

$$\frac{dy}{dx} = \frac{v}{u} = -\frac{y}{x}$$

or

$$\frac{dy}{y} = -\frac{dx}{x},$$

and therefore

$$\ln y = \ln C - \ln x$$

or

$$xy = \text{const.}$$

The streamlines form a family of rectangular hyperbolas whose asymptotes are the x- and y-axes (Fig. 72). These streamlines may be considered as those of a two-dimensional motion against a wall. In many cases such streamline motion occurs as a part of a more general flow. It is then possible to expand this more general potential function in the neighborhood of the point of bifurcation into a power series. If all terms of higher degree than the second of this series are neglected, we obtain the above Φ-function, which describes the field in a small area near the point of bifurcation.

In the case of steady motion the distribution of pressure is given by the equation

$$\frac{\mathbf{w}^2}{2} + \frac{p}{\rho} = \frac{a^2}{2}(x^2 + y^2) + \frac{p}{\rho} = C.$$

Denoting the pressure at the stagnation point by p_1 and taking this point as origin, we have

$$0 + \frac{p_1}{\rho} = C;$$

and, since the constant is equal to that of the previous equation,

$$p_1 - p = \rho \frac{a^2}{2}(x^2 + y^2).$$

The curves of equal pressure are therefore a family of concentric circles.

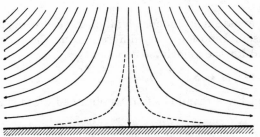

Fig. 72.—Two-dimensional fluid motion against a plate.

(b) As a second solution of the equation $a + b + c = 0$ we take

$$b = a, \qquad c = -2a,$$

so that

$$\Phi = \frac{a}{2}(x^2 + y^2 - 2z^2).$$

The velocity components are now

$$u = ax, \qquad v = ay, \qquad w = -2az.$$

The equations for the streamlines are

$$dx : dy : dz = x : y : -2z,$$

so that the projection of the streamlines on the (x, y)-plane is

$$\frac{dx}{dy} = \frac{x}{y}$$

or

$$\frac{dx}{x} = \frac{dy}{y}.$$

Therefore

$$\ln x = \ln y + \ln C$$

or

$$x = Cy.$$

The projection of the streamlines on the xy-plane is a family of straight lines through the origin.

The projection on the xz-plane is given by

$$\frac{dx}{dz} = -\frac{x}{2z}$$

or

$$\frac{dx}{x} = -\frac{dz}{2z}.$$

Integrated:

$$\ln x = \ln C - \tfrac{1}{2} \ln z$$

or

$$x^2 z = \text{const.},$$

so that the projection of the streamlines on the xz-plane is a family of cubic hyperbolas with the x- and z-axes as asymptotes.

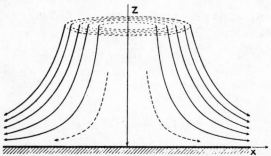

FIG. 73.—Three-dimensional fluid motion against a plate (jet).

Figure 73 shows the form of the streamlines which are to be considered rotationally symmetrical about the z-axis.

Since $\mathbf{w} = 0$ for $z = 0$, the xy-plane is a possible bounding surface to the motion, which may be described as that of a wide jet, symmetrical about the center line impinging against an infinite plate. The center filament of the jet meets the plate in a stagnation point.

In this case, a may be a function of the time, but, since the streamline equations do not involve a, the streamlines, path lines, and streak lines are all identical, as in the previous case. The time affects only the numerical value of the velocity but not its direction.

For steady motion the distribution of the pressure is given by

$$\frac{p}{\rho} + \frac{a^2}{2}(x^2 + y^2 - 4z^2) = \text{const.}$$

The surfaces of equal pressure form a family of ellipsoids whose axes bear the ratio $1:1:\frac{1}{2}$ to each other.

If a is a function of t, the general Bernoulli equation for the pressure distribution is

$$\frac{p}{\rho} = f(t) - \frac{1}{2}\left[\frac{da}{dt}(x^2 + y^2 - 2z^2) + a^2(x^2 + y^2 + 4z^2)\right],$$

where $f(t)$ is defined if the pressure variation at any one point of the fluid is known. The shape of the surfaces of equal pressure depends on the value of the ratio of a^2 to da/dt; for example, when $da/dt = -a^2$,

$$\frac{p}{\rho} = f(t) - 3a^2z^2,$$

p is constant on planes parallel to the (x, y)-plane; when $da/dt = a^2$,

$$\frac{p}{\rho} = f(t) - a^2(x^2 + y^2 + z^2),$$

p is constant on spheres; when $da/dt = 2a^2$,

$$\frac{p}{\rho} = f(t) - \frac{3}{2}a^2(x^2 + y^2),$$

p is constant on circular cylinders whose axes are parallel to the z-axis.

68. The Source and Sink Potential.—Let us consider the function $\Phi = b/r$, where $r = \sqrt{x^2 + y^2 + z^2}$ is the distance from the origin of coordinates. Since the surfaces $\Phi = \text{constant}$ are spheres concentric with the origin, and since the velocity $\mathbf{w} = \text{grad } \Phi$ is perpendicular to the surface $\Phi = \text{constant}$, the motion is purely radial with the velocity w_r, where

$$w_r = \frac{\partial \Phi}{\partial r} = -\frac{b}{r^2}.$$

The question whether the function $\Phi = b/r$ satisfies the equation $\Delta\Phi = 0$ has the same answer as the question whether the equation of continuity is or is not satisfied, since that equation is also $\Delta\Phi = 0$. If Q be the mass of liquid flowing through a sphere of radius r in unit time, then

$$Q = 4\pi r^2 w_r = -4\pi b.$$

Since this expression is independent of r, it is clear that the same quantity of fluid passes through every concentric spherical surface. If we consider a shell bounded by two spheres r_1 and r_2, we see that the amount entering the larger sphere in unit time is exactly equal to the amount leaving the smaller sphere in the same time. The equation of continuity is thus satisfied. It also can easily be verified, however, that the sum of the three partial second differentials of Φ is identically zero.

When b is positive we have a radial motion directed toward the origin, and when it is negative the motion is in the outward direction.

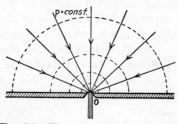

FIG. 74.—Flow through a small hole in a flat plate. (Sink motion.)

On approaching the origin O, the velocities increase inversely proportional to the square of the distance from the origin. The origin itself is a singularity in that its velocity is infinite and in all directions at once. Moreover, if b is positive, a quantity of fluid $Q = 4\pi b$ is annihilated each second in the origin; and if b is negative, a similar amount of fluid is generated there. Physically this is impossible, and the potential function $\Phi = b/r$ only expresses motion outside the origin O.

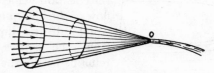

FIG. 75.—Flow through a cone.

If b is positive, the point O is called a "sink," and when it is negative we speak of it as a "source." There are many applications of this particular potential function. For example, it expresses roughly the motion through a small hole in a large plate (Fig. 74) at least in a region sufficiently far away from the hole. The hole acts as if it were a sink; the pressure decreases very rapidly on approaching it. Since the pressure changes as the square of the velocity, and the velocity itself is proportional to the inverse square of the distance from the hole, the pressure decreases toward the origin proportional to the inverse fourth power of the distance. By halving our distance from the

hole we experience a sucking action 16 times as strong. Excluding the immediate neighborhood of the hole, the streamlines are radial so that the boundary condition of zero velocity normal to the plate is satisfied.

The motion can also represent the flow through a cone (Fig. 75). Since the fluid masses flowing through all cross sections per unit time must be identical, it follows that the velocity must be inversely proportional to the square of the distance from the apex of the cone.

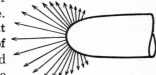

Fig. 76.—Stream lines at the blunt nose of a body moving through still fluid. (Non-steady motion.)

69. Description of the Motion about a Body of Revolution by the Method of Sources and Sinks.—We now proceed to a further application of the source and sink potential which is due to Rankine.

The round-nosed body, shown in Fig. 76, pushes all fluid particles aside during its motion, just as if it carried a source in its interior. Now bring the body to rest by superposing on the system a uniform motion to the right ($\Phi = ax$). This transforms the picture of Fig. 76 into that of Fig. 77, in which the stream splits into two parts.

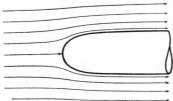

Fig. 77.—Flow round the blunt nose of a body at rest. (Steady motion.)

On closer consideration it is evident that the superposition of the constant velocity on the source flow of the body builds a system in which all the liquid emanating from the source in the body remains within the body and flows off to the right (Fig. 78). If we then insert a sink in the rear portion of the body, such that the water produced by the source is completely sucked in, there remains only the deformation in the streamlines exterior to the body. In this way it is possible, by the help of potential functions, to calculate the flow round a body of revolution. In addition to a single point source and sink in the body we can assume many more of them or even continuous (or discontinuous) distributions of sources and sinks.

By making different assumptions about the distribution of sources and sinks along the axis of symmetry of the body a great variety of shapes can be obtained. It is only essential that the sinks destroy as much liquid as the sources produce.

As an example we shall consider in closer detail the case of a point source. The potential function derived by superposing the uniform flow $\Phi_1 = ax$ on the source motion $\Phi_2 = -b/r$ is:

$$\Phi = \Phi_1 + \Phi_2 = ax - \frac{b}{r},$$

where

$$r = \sqrt{x^2 + y^2 + z^2}.$$

On the axis of symmetry, *i.e.*, $y = z = 0$, when x is negative $(r = -x)$, we have

$$\Phi_2 = \frac{b}{x};$$

and therefore

$$\Phi = ax + \frac{b}{x},$$

so that

$$u = \frac{\partial \Phi}{\partial x} = a - \frac{b}{x^2}.$$

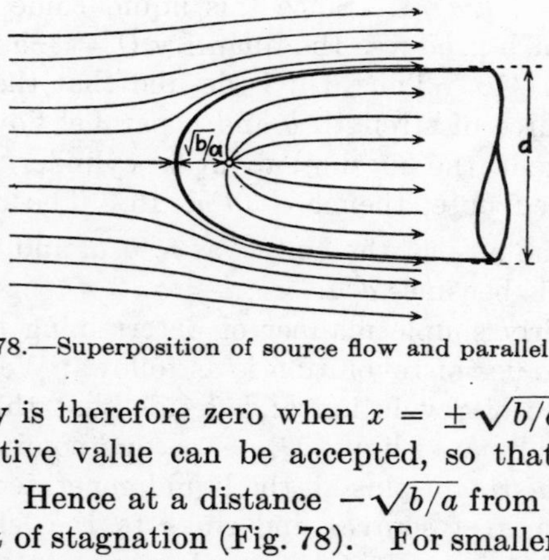

FIG. 78.—Superposition of source flow and parallel flow.

The velocity is therefore zero when $x = \pm\sqrt{b/a}$. Obviously only the negative value can be accepted, so that $u = 0$ when $x = -\sqrt{b/a}$. Hence at a distance $-\sqrt{b/a}$ from the source we have the point of stagnation (Fig. 78). For smaller values of $-x$ the velocity is toward the left; here the velocities due to the source are larger than those due to the uniform stream, but, as $-x$ increases, the value of the source velocity decreases to zero with the inverse square of the distance, so that at large distances we have only the parallel flow.

When x is positive we must write $r = +x$, and therefore

$$\Phi = ax - \frac{b}{x}$$

and

$$u = \frac{\partial \Phi}{\partial x} = a + \frac{b}{x^2}.$$

To the right of the source the velocity is thus always directed to the right and approaches the value $u = a$ for $x = \infty$.

The individual streamlines due to the potential $\Phi = ax - b/r$ can be calculated from the general formula $dx:dy:dz = \quad :v:w$. The streamline passing through the stagnation point

$$x = -\sqrt{\frac{b}{a}},$$

when revolved round the x-axis, obviously gives the surface separating the general left to right flow from the source-sink flow. This surface of revolution therefore behaves as a solid body in a stream flowing to the right with the general velocity a.

If d be the diameter of this body at a sufficient distance from the source, where it is nearly cylindrical, then the amount of fluid Q flowing to the right through the cross section per second is $Q = (\pi d^2 a)/4$. Since this liquid comes entirely from the source which produces the quantity $Q = 4\pi b$ per second, we have $b = (d^2 a)/16$. Thus if it is desired that the superposition of a point source of strength b and a parallel flow of velocity a should represent the motion about a cylinder of diameter d with a rounded nose, then $b = (d^2 a)/16$. The distance of the source from the nose of the body was $\sqrt{b/a}$, and with the above value for b this becomes $d/4$.

A particularly simple manner of determining the streamlines round such bodies of revolution is as follows: We consider some stream surface of revolution and decide that this surface shall transport a volume of liquid Q_1 per second coming from the left with the main stream plus all the liquid generated per second by the combination of sources and sinks to the left of the point under consideration. If Q_1 is made equal to a constant, we obviously remain on the same stream surface. By introducing cylindrical coordinates (x and r, where $r = \sqrt{x^2 + y^2}$), the following relation must be true for every section perpendicular to the axis of symmetry:

$$\int_0^{r_1} 2\pi r u \, dr = Q_1 + Q_2,$$

where Q_2 is the algebraic sum of the volumes of liquid issuing per second from the sources and sinks to the left of the section,

and r_1 lies on some streamline (or surface). Different values of Q_1 produce different streamlines. If Q_1 is zero, we have the special surface of revolution which separates the main stream from the source-sink stream, *i.e.*, the surface of the rigid body itself.

This source-sink method has been further developed, particularly for airship bodies, by G. Fuhrmann.[1] In his work a number of surfaces of revolution were investigated by assuming different combinations of sources and sinks. The form of the streamlines and the distribution of pressure were calculated and were compared to the values derived from experiments.

70. The Motion about a Sphere; Doublets.—Let the distance between source and sink gradually become smaller, then the form of the body round which the flow takes place also becomes less elongated and, moreover, its general size diminishes. When the intensity of the source and sink (which are equal) is kept constant, the dimensions of the body decrease proportionately with the distance between the source and sink, until finally both become zero together.

If in the limit we desire to obtain a body of finite size, we must increase the intensity of the sink and source in the same ratio as the distance between them decreases. If the distance between the source ($\Phi = -b/r_1$) and the sink ($\Phi = b/r_2$) is decreased to Δx, then the expression for the combination with the required increasing intensity is

$$\Phi = \frac{-\dfrac{b}{r_1} + \dfrac{b}{r_2}}{\Delta x};$$

or, with the notations

$$\frac{b}{r_1} = \Phi_1, \qquad \frac{b}{r_2} = \Phi_2,$$

this can also be written as

$$\Phi = \frac{\Phi_2 - \Phi_1}{\Delta x};$$

but, since $\Phi_2(x, y, z) = \Phi_1(x - \Delta x, y, z)$, we have in the limit as source and sink approach:

[1] FUHRMANN, G., Theoretical and Experimental Investigations on Balloon Models (German), Dissertation, Göttingen, 1912, *Jahrb. Motorluftschiffstudiengesellschaft*, vol. 5, p. 63, 1911–1912.

$$\Phi = \lim_{\Delta x = 0} \frac{\Phi_1(x - \Delta x, y, z) - \Phi_1(x, y, z)}{\Delta x} = -\frac{\partial \Phi_1}{\partial x}.$$

This gives, since Φ is dependent only on r,

$$\Phi = -\frac{\partial \Phi_1}{\partial x} = \frac{\partial \Phi_1}{\partial r} \cdot \frac{\partial r}{\partial x} = \frac{b}{r^2} \cdot \frac{\partial r}{\partial x}.$$

With

$$r = \sqrt{x^2 + y^2 + z^2}$$

and, consequently,

$$\frac{\partial r}{\partial x} = \frac{x}{\sqrt{x^2 + y^2 + z^2}} = \frac{x}{r},$$

this becomes

$$\Phi = \frac{b}{r^2} \cdot \frac{x}{r} = \frac{b}{r^2} \cos \varphi,$$

where x is written as $r \cos \varphi$. We have expressed Φ in polar coordinates r and φ. Φ is called the "potential function of a doublet."

The radial velocity in the direction φ is

$$w_r = \left(\frac{\partial \Phi}{\partial r}\right)_\varphi = -\frac{2b}{r^3} \cos \varphi.$$

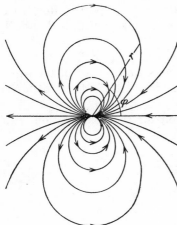

If we consider a sphere of radius r_1, the radial velocity on its surface is $w_r =$ const. $\cos \varphi$, and therefore, when φ lies between 0 and 90 deg (and also between 270 and 360 deg), the radial velocity is negative; but it is positive when it lies between 90 and 180 deg or between 180 and 270 deg (see Fig. 79).

FIG. 79.—Dipole or doublet.

If on the other hand we impart a velocity $-a$ (parallel to the x-axis in the negative direction) to the sphere, then obviously the normal components of the velocity of the fluid on the surface of the sphere must be equal to that of the sphere itself, *i.e.*, $-a \cos \varphi$ (Fig. 80). The formula for the radial velocity at a distance r_1 from the origin, derived from the potential of a doublet was

$$w_r = -\frac{2b}{r_1{}^3} \cos \varphi = \text{const. } \cos \varphi.$$

Hence the velocity field of our potential $\Phi = b/r^2 \cos \varphi$ is identical with that of the moving sphere (since the boundary condition is satisfied), when only the velocity of the sphere $-a$ is made equal to $-2b/r_1^3$. Since the constant b is still at our disposal we can make it $b = ar_1^3/2$, so that

$$\Phi = \frac{1}{2}a\frac{r_1^3}{r^2} \cos \varphi$$

is the potential of the fluid motion produced when a sphere of radius r_1 moves with a uniform velocity $-a$. As we have seen, the velocity in some direction φ decreases very rapidly, namely, with the third power of the distance from the center.

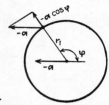

Fig. 80.—Motion of a sphere with velocity a from right to left.

The potential function of the motion around a sphere at rest can be derived from the last result by superposing a velocity a to the right on the whole system, *i.e.*, by adding a term $\Phi = ax$ to the potential, so that

$$\Phi = a\left(x + \frac{r_1^3}{2r^2} \cos \varphi \right);$$

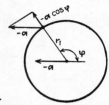

Fig. 81.—Motion round a sphere.

or with $x = r \cos \varphi$:

$$\Phi = a\left(r + \frac{r_1^3}{2r^2} \right) \cos \varphi. \tag{9}$$

The streamlines are shown in Fig. 81.

Since the sphere is now at rest relative to the frame of reference, the radial velocity on the surface of the sphere, $r = r_1$, must be zero. This agrees with the last equation

$$\left(\frac{\partial \Phi}{\partial r} \right)_{r=r_1} = (w_r)_{r_1} = 0.$$

At a great distance from the sphere the second term $r_1^3/2r^2$ is of no importance and we are left with only the potential

$ar \cos \varphi = ax$ which represents parallel flow. The motion in the interior of the sphere, which is derived from the analytic expression, is shown in Fig. 82. Here the second term of Eq. (9) is the more important one.

We have already noted in Art. 55 that the actual motion of a fluid about a sphere (in the steady state) is quite different,

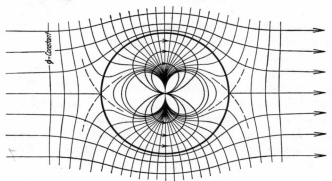

Fig. 82.—Superposition of a dipole flow and a parallel flow from left to right.

even when the viscosity is very small, and that the form of the streamlines is altered by that part of the boundary layer which detaches itself from the sphere. However at the first instant after starting from rest the streamlines actually conform to those of Fig. 81, and the potential is that given in Eq. (9).

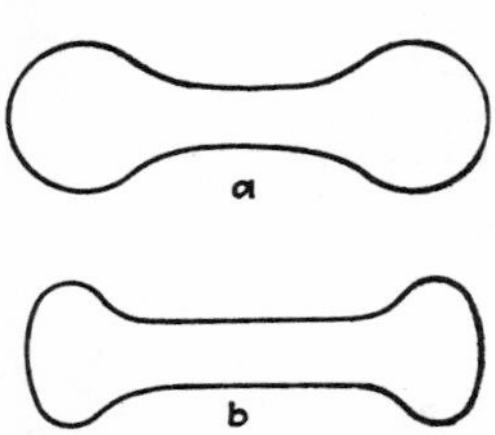

By combining sources and sinks in different ways it is possible to build up bodies of revolution of a great variety of shapes including the one shown in Fig. 83a. The same cannot be said, however, of the body shown in Fig. 83b. Here the curvature is too small at the stagnation point and the width of the thick ends is too large. In general, elongated bodies can be represented by sources and sinks with the sphere as the boundary case which just can be treated in this manner by bringing the source and sink infinitely close together.

Fig. 83.—(a) Axisymmetric body which can still be treated by the method of sources and sinks. (b) Axisymmetric body for which this is no longer possible.

The calculation of the potential for bodies of the form 83b can be achieved by introducing vortex rings in addition to sources and sinks.

71. The Potential of a Rectilinear Vortex.—Another simple example of potential motion is given by the equation

$$\Phi = c\varphi,$$

where φ is the angle between a plane containing the z-axis, and the x,z-plane. The potential is independent of z and thus is the same in all planes parallel to the x,y-plane. This is an example of a plane or two-dimensional problem, which will be investigated in detail in the next chapter.

The surfaces of constant potential form a pencil of planes, whose intersections with the x,y-plane form a pencil of rays through the origin. The streamlines are circles in planes at right angles to the z-axis with their centers on the axis; the radial velocity w_r is zero, the velocity perpendicular to the radius is

$$w_\varphi = \frac{\partial \Phi}{r \partial \varphi} = \frac{c}{r}$$

It is immediately evident from physical considerations that such a type of motion satisfies the equation of continuity. Moreover we can substitute $\varphi = \tan^{-1} y/x$ into the potential $\Phi = c\varphi$ and then calculate $\Delta\Phi$, which will be found zero. The function $\Phi = c\varphi$ is therefore indeed a potential function, though it has a singularity at the origin.

This form of motion is called a "vortex." The velocity at every point is perpendicular to the radius vector, and its value is inversely proportional to the distance from the origin.

The potential $\Phi = c\varphi$ has a property which is worthy of note: As φ increases from zero, the potential also increases till at $\varphi = 2\pi$ it has the value $2\pi c$. Another circuit round the origin increases Φ by another $2\pi c$. We are dealing with a so-called many-valued potential function which confronts us with some new difficulties. The general Bernoulli equation on account of the term $\partial\Phi/\partial t$ would imply a multiplicity of values for the pressure, which is physically impossible. We must therefore conclude that a motion given by the many-valued potential $\Phi = c\varphi$ can only persist if at some instant it already exists, but it cannot be produced from rest.

There is a way in which a vortex can be generated; however, then, the continuity of the fluid has to be temporarily destroyed by the insertion of a rigid body. Consider a thin plate accelerated rapidly over a short distance, where p_1 and p_2 are the

pressures on both sides of the plate and $p_1 > p_2$. This pressure difference produces a streamline pattern shown in Fig. 84. Now imagine the plate suddenly removed in a direction at right angles to the fluid surface, the pressure evens out and as an approximation to the actual motion we have a family of concentric circles.

The line integral of the velocity due to the potential $\Phi = c\varphi$

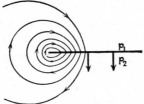

round a curve which surrounds the singular point at the origin is

$$\oint \mathbf{w} \circ d\mathbf{r} = \Phi(2\pi) - \Phi(0) = 2\pi c.$$

This expression, which we denote by Γ, is called the "circulation."

FIG. 84.—Formation of a vortex in a non-viscous fluid.

72. Difference between Potential Motion with Circulation and a Motion with Rotation.—How can we say that the fluid motion round a vortex considered in the previous section is a potential motion, when potential motion implies freedom from "rotation"? In explaining this we remember that the definition of the term "rotation," as given in Art. 45, refers to a property of the individual fluid particles, and defined in this way the motion discussed in Art. 71 really is irrotational, except for the singular point $r = 0$. In order to realize this, consider a small fluid "rod" in the position shown in Fig. 85a. In an interval of time dt, the fluid particle moves forward a distance $w_\varphi dt$ and rotates through an angle $d\varphi$ in a counterclockwise direction where

$$d\varphi = \frac{w_\varphi dt}{r} = \frac{c\,dt}{r^2}.$$

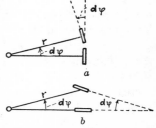

FIG. 85a–b.—Motion of a "fluid rod" in an irrotational velocity field.

If, however, we follow the motion of a rod in the position of Fig. 85b, the rod turns in a clockwise sense, because the velocity near the origin is greater than farther away from it. It turns through an angle $d\varphi'$, where

$$d\varphi' = \frac{\dfrac{\partial w_\varphi}{\partial r}dr\,dt}{dr} = \frac{\partial w_\varphi}{\partial r}dt = -\frac{c}{r^2}dt.$$

We therefore see that when one axis of a fluid particle turns through a certain angle $d\varphi$, the axis at right angles to it turns through the same angle in the opposite direction so that the resulting mean value for the rotation is zero:

$$\tfrac{1}{2}(d\varphi + d\varphi') = 0.$$

If in Fig. 86 two dotted lines are drawn at angles of 45 deg to the axes of the liquid "cross," the direction of these lines remains unchanged after a small motion $d\varphi$. We therefore see that the fluid element has two axes which remain parallel to their original direction even after the deformation. This means that the motion is irrotational.

If one forms the line integral $\oint \mathbf{w} \circ d\mathbf{r}$ over a closed curve which does not surround the singular point, it is everywhere zero. If the curve along which the integral is taken surrounds the singular point, its value is $2\pi c$. The integral along a closed curve in potential motion is zero and only differs from zero when it has a singular point within it. On the other hand with a rotational motion the line integral generally is different from zero, no matter around what curve it is taken.

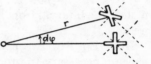

FIG. 86.—Motion of a "fluid cross" in an irrotational velocity field.

The value of the line integral gives a measure for the intensity of the rotation as we have seen in Art. 45.

73. Interpretation of Potential as Impulsive Pressure.—The general Bernoulli equation for non-steady motion of an incompressible fluid is

$$\frac{\partial \Phi}{\partial t} + \frac{\mathbf{w}^2}{2} + \frac{p}{\rho} - U = f(t).$$

If now we assume that the motion represented by this equation is produced from rest by some sort of impulsive pressure, then in the first instant of motion the acceleration $\partial \mathbf{w}/\partial t$ is large in comparison with the terms $\mathbf{w} \circ \nabla \mathbf{w}$, and likewise the gradient of the impulsive pressure is large in comparison with the action of gravity. The value of $\partial \Phi/\partial t$ therefore swamps those of $\mathbf{w}^2/2$ and U in the general Bernoulli equation. We now form the time integral over the duration of the impulse τ and neglect $\mathbf{w}^2/2$ and U, obtaining thereby

$$\int_0^\tau \left\{ \frac{\partial \Phi}{\partial t} + \frac{p}{\rho} - f(t) \right\} dt = 0.$$

We know however that $f(t)$ only means that the pressure at some fixed point has a definitely prescribed value, and therefore

$$\int_0^\tau f(t) dt = \text{const.}$$

Further,

$$\int_0^\tau \frac{\partial \Phi}{\partial t} = \Phi(\tau) - \Phi_0.$$

But according to our assumption the fluid was at rest prior to the impulse, so that $\Phi_0 = \text{constant}$, and we therefore have

$$\Phi(\tau) + \frac{1}{\rho} \int_0^\tau p \, dt = \text{const.} \tag{10}$$

The integral $\dfrac{1}{\rho} \displaystyle\int_0^\tau p \, dt$ represents the general effect of the pressure in the time interval from 0 to τ and is a measure of the "impulsive pressure."

From Eq. (10) we can calculate the value of the impulsive pressure necessary to produce a motion given by the potential Φ. Since the pressure is continuously distributed over space, the potential must be similarly distributed, *i.e.*, impulsive pressure action cannot produce a discontinuous potential. Since p can have only *one* value at each point, Φ must have the same property, *i.e.*, Φ is also unique, which means an impulsive pressure cannot generate a motion with circulation.

Equation (10) now gives a simple significance to the potential function. In any uniform potential motion we may consider the product of the density and the potential as the impulsive pressure necessary to generate this motion from rest. We can also assign to the potential a similar meaning when dealing with jets, *i.e.*, with discontinuous potentials, in which cases rigid walls have to be assumed in place of the surfaces of discontinuity.

CHAPTER XI

TWO-DIMENSIONAL POTENTIAL MOTION

74. The Real and Imaginary Parts of an Analytic Function of Complex Argument Are Solutions of Laplace's Differential Equation.—Although exact examples of two-dimensional flow hardly ever occur, there are many cases where the motion, either wholly or in part, can be considered as a good approximation to it. The particular importance of the two-dimensional problem lies in the fact that it is particularly amenable to mathematical analysis.

The simplification in the mathematical treatment is not essentially caused by having only two variables instead of three (this simplification also exists in cases of rotational symmetry) but is rather connected with the fact that, as soon as the analysis depends on the two Cartesian coordinates (x, y), the real and imaginary parts of any analytic function of the complex argument $(x + iy)$ satisfies the Laplace differential equation of potential theory.

The analytic function $F(x + iy) = F(z)$ of the complex argument $(x + iy) = z$ can always be split up into a real and an imaginary part:

$$F(z) = F(x + iy) = \Phi(x, y) + i\Psi(x, y),$$

where Φ and Ψ are real functions of x and y.

Let us form the second partial derivatives of $F(z)$ with respect to x and y. Remembering that

$$\frac{\partial z}{\partial x} = 1,$$

$$\frac{\partial z}{\partial y} = i,$$

$$\frac{\partial^2 z}{\partial x^2} = \frac{\partial^2 z}{\partial y^2} = 0,$$

we find

$$\frac{\partial^2 F}{\partial x^2} = \frac{d^2 F}{dz^2}\left(\frac{\partial z}{\partial x}\right)^2 = \frac{d^2 F}{dz^2} \cdot 1$$

and

$$\frac{\partial^2 F}{\partial y^2} = \frac{d^2 F}{dz^2}\left(\frac{\partial z}{\partial y}\right)^2 = \frac{d^2 F}{dz^2} \cdot (-1),$$

so that

$$\frac{\partial^2 F}{\partial x^2} + \frac{\partial^2 F}{\partial y^2} = 0.$$

With $F = \Phi + i\Psi$, this becomes

$$\frac{\partial^2 \Phi}{\partial x^2} + \frac{\partial^2 \Phi}{\partial y^2} + i\left(\frac{\partial^2 \Psi}{\partial x^2} + \frac{\partial^2 \Psi}{\partial y^2}\right) = 0,$$

and, therefore,

$$\frac{\partial^2 \Phi}{\partial x^2} + \frac{\partial^2 \Phi}{\partial y^2} = 0, \qquad \frac{\partial^2 \Psi}{\partial x^2} + \frac{\partial^2 \Psi}{\partial y^2} = 0.$$

Thus both the real and imaginary parts of each analytic function of the complex argument $(x + iy) = z$ satisfy the Laplace differential equation.

75. The Cauchy-Riemann Differential Equations and Their Physical Interpretation.

From

$$F(x + iy) = \Phi(x, y) + i\Psi(x, y),$$

we find

$$\frac{\partial F}{\partial x} = \frac{dF}{dz} \cdot \frac{\partial z}{\partial x} = \frac{dF}{dz}$$

and

$$\frac{\partial F}{\partial y} = \frac{dF}{dz} \cdot \frac{\partial z}{\partial y} = i\frac{dF}{dz},$$

thus giving

$$\frac{\partial F}{\partial y} = i\frac{\partial F}{\partial x}.$$

But, since

$$\frac{\partial F}{\partial x} = \frac{\partial \Phi}{\partial x} + i\frac{\partial \Psi}{\partial x}$$

and

$$\frac{\partial F}{\partial y} = \frac{\partial \Phi}{\partial y} + i\frac{\partial \Psi}{\partial y},$$

we have

$$\frac{\partial \Phi}{\partial y} + i\frac{\partial \Psi}{\partial y} = i\frac{\partial \Phi}{\partial x} - \frac{\partial \Psi}{\partial x};$$

and, separating real and imaginary parts,

$$\left.\begin{aligned}\frac{\partial \Phi}{\partial x} &= \frac{\partial \Psi}{\partial y} \\ \frac{\partial \Phi}{\partial y} &= -\frac{\partial \Psi}{\partial x}\end{aligned}\right\} . \tag{1}$$

These are the so-called "Cauchy-Riemann differential equations" and they follow immediately from the assumption that Φ and Ψ are the real and imaginary parts of an analytic function of the complex argument $x + iy$. If we differentiate these equations partially with respect to x and y, we again arrive at Laplace's equation.

The physical meaning of the Cauchy-Riemann equations can be understood as follows: If Φ is the potential function, then

$$u = \frac{\partial \Phi}{\partial x}$$

and

$$v = \frac{\partial \Phi}{\partial y}.$$

By Eq. (1) we also have

$$u = \frac{\partial \Psi}{\partial y}$$

and

$$v = -\frac{\partial \Psi}{\partial x}.$$

If we form the gradient of Φ:

$$\text{grad } \Phi = \mathbf{i}\frac{\partial \Phi}{\partial x} + \mathbf{j}\frac{\partial \Phi}{\partial y} = \mathbf{i}u + \mathbf{j}v$$

($\mathbf{i}$, $\mathbf{j}$ are the unit vectors), and also the gradient of Ψ,

$$\text{grad } \Psi = \mathbf{i}\frac{\partial \Psi}{\partial x} + \mathbf{j}\frac{\partial \Psi}{\partial y} = -\mathbf{i}v + \mathbf{j}u,$$

we see that

$$\text{grad } \Phi \perp \text{grad } \Psi$$

and

$$|\text{grad } \Phi| = |\text{grad } \Psi|,$$

as is shown in Fig. 87. It follows that the families of curves $\Phi = \text{constant}$ and $\Psi = \text{constant}$ form an orthogonal system and that when the intervals between consecutive values of Φ and Ψ are equal and sufficiently small, they form a network of squares.

Now, since $\mathbf{i}u + \mathbf{j}v = \mathbf{w}$, *i.e.*, since the direction of the velocities (the streamlines) are perpendicular to the curves $\Phi = \text{constant}$, and since the curves $\Psi = \text{constant}$ are also perpendicular to the curves $\Phi = \text{constant}$, it follows that the set of curves $\Psi = \text{constant}$ forms the family of streamlines. When some of the curves

Φ = constant and Ψ = constant are known, it is possible to find new members of the families by interpolation, owing to the fact that these two sets form a network of squares. If the diagonal curves of some squares of the net are produced, as

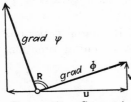

F I G . 8 7 .—Geometrical relation between the gradient of the potential function and that of the stream function.

in Fig. 88, then the points of intersection of these diagonals are also points on the curves Φ = constant and Ψ = constant.

It should be noted that not only Φ but also Ψ can be considered as the potential of some motion, because $F^*(z) = -iF(z)$ is also an analytic function. In fact it was already seen in Art. 74 that $\Delta\Psi = 0$. In this case the lines Φ = constant are the streamlines.

We shall call the function

$$F(z) = F(x + iy) = \Phi(xy) + i\Psi(xy)$$

the stream function.

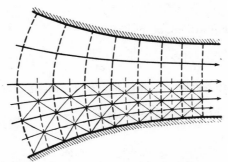

F ɪ ɢ. 88.—The construction of further streamlines when some of the curves Φ = constant, ψ = constant, are given.

The total differential dF of $F(x + iy)$ is

$$\begin{aligned}
dF &= \frac{\partial F}{\partial x}dx + \frac{\partial F}{\partial y}dy \\
&= \left(\frac{\partial \Phi}{\partial x} + i\frac{\partial \Psi}{\partial x}\right)dx + \left(\frac{\partial \Phi}{\partial y} + i\frac{\partial \Psi}{\partial y}\right)dy \\
&= (u - iv)dx + (v + iu)dy \\
&= (u - iv)dx + i(u - iv)dy \\
&= (u - iv)(dx + idy) \\
&= \overline{w}\,dz,
\end{aligned}$$

where $\overline{\mathbf{w}}$ is the conjugate complex to $\mathbf{w} = u + iv$, which can be obtained by reflecting $\mathbf{w}$ in the real axis. Therefore we have

$$\overline{\mathbf{w}} = \frac{dF(z)}{dz} = F'(z).$$

This simple relation states that the derivative of the stream function $F(z)$ with respect to the complex argument z is equal to the reflection of the velocity vector in the real axis.

76. The Stream Function.—We shall now explain in another way the meaning of Ψ and the fact that the equation $\Psi = $ constant represents the family of streamlines.

In the flow along a wall, as shown in Fig. 89, it is of interest to know what quantity of liquid Q flows between the wall and the

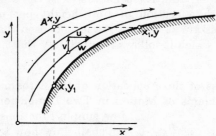

Fig. 89.—Two-dimensional flow along a wall.

point A (of coordinates x and y) per unit time. The amount flowing across a rectangle made up of a line drawn from A parallel to the x-axis and of depth h perpendicular to the paper is

$$Q = h \int_x^{x_1} v\, dx;$$

and with

$$v = -\frac{\partial \Psi}{\partial x},$$

this becomes

$$Q = h(\Psi - \Psi_1).$$

Now we draw a line through A parallel to the y-axis, and we find similarly

$$Q = h \int_{y_1}^y u\, dy,$$

but

$$u = \frac{\partial \Psi}{\partial y},$$

so that again

$$Q = h(\Psi - \Psi_1),$$

from which

$$\Psi = \frac{Q}{h} + \text{const.}$$

The curves $\Psi = \text{constant}$ are obviously curves which between themselves and the boundary let through constant amounts of fluid per unit time, *i.e.*, they are streamlines. Ψ is therefore called the "stream function."

If h, the length in the direction of the z-axis, is taken equal to unity, then

$$\Psi = Q + \text{const.}$$

Hence, but for an additive constant, Ψ is the amount of liquid flowing through a section of unit height per unit time. In general the constant of integration is so chosen that Ψ is zero on the wall, from which it follows immediately that

$$\Psi = Q.$$

77. Examples of the Application of the Stream Function $F(z)$ to Simple Problems of Motion in Two Dimensions.—Taking a few simple functions as examples, we shall show how to calculate the streamline pattern from a given analytic function of a complex variable, and we shall see how in many cases this can be done quite easily.

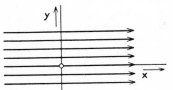

Case 1.

Fig. 90.—Two-dimensional flow: Stream function $F(z) = az$ (a real).

$$F = az.$$

If a is real, then

$$\Phi = ax \quad \text{and} \quad \Psi = ay.$$

The streamlines $\Psi = \text{constant}$ are therefore parallel to the x-axis, and the curves of equal potential are parallel to the y-axis. The reflection of the velocity vector in the real axis is

$$\overline{w} = F'(z) = a = u - iv,$$

so that

$$u = a, \quad v = 0,$$

which means that we have a flow parallel to the axis of x (Fig. 90).

If a is complex and equal to $a_1 + ia_2$, then

$$F(z) = (a_1 + ia_2)(x + iy) = \underbrace{a_1 x - a_2 y}_{\Phi} + i\underbrace{(a_1 y + a_2 x)}_{\Psi}$$

and the conjugate complex velocity vector is

$$\overline{w} = F'(z) = a_1 + ia_2 = u - iv,$$

so that

$$u = a_1, \qquad v = -a_2.$$

We see that since $u/v = -a_1/a_2 = $ constant, we are dealing with a rectilinear flow (Fig. 91).

All linear functions of z give straight-line motions. In the following examples a will be considered as real. This does not lessen the generality, since making a complex only introduces a rotation of the whole streamline pattern through some angle.

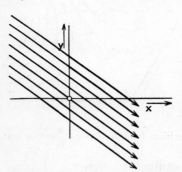
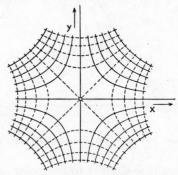

FIG. 91.—Two-dimensional flow: Stream function $F(z) = az$ (a complex).

FIG. 92.—Two-dimensional flow: Stream function $F(z) = az^2/2$ (a real).

Case 2.

$$F = \frac{a}{2}z^2 = \underbrace{\frac{a}{2}(x^2 - y^2)}_{\Phi} + \underbrace{iaxy}_{\Psi}.$$

The streamlines $\Psi = axy = $ const. give a series of rectangular hyperbolas with the coordinate axes as asymptotes, while $\Phi = \frac{a}{2}(x^2 - y^2) = $ const. gives the family of orthogonal rectangular hyperbolas whose asymptotes are the lines $y = x$, and $y = -x$. The mesh of squares between the curves $\Phi = $ const. and $\Psi = $ const. is shown in Fig. 92.

Parts of this diagram can sometimes be of practical significance; one-half of it for example depicts the case of the flow against a plate (Fig. 72); one-quarter of it is the flow inside a corner.

The velocity $\overline{w} = u - iv$ is obtained by differentiating the function with respect to z:

$$\overline{w} = F'(z) = az, \qquad u = ax, \qquad v = -ay.$$

The velocity is therefore proportional to the complex quantity z. Hence, if we have a series of concentric circles with the origin as center, the velocity at any point is proportional to the radius of the circle through that point.

Case 3.

$$F = \frac{a}{n}z^n.$$

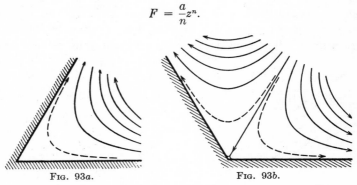

Fɪɢ. 93a.   Fɪɢ. 93b.

Fɪɢs. 93*a–b*.—Two-dimensional flow: Stream function $F(z) = \dfrac{a}{3} az^3$ (*a* real).

Here it is simpler to use polar coordinates $z = re^{i\varphi}$:

$$F = \frac{a}{n}r^n e^{in\varphi} = \frac{a}{n}r^n(\cos n\varphi + i \sin n\varphi)$$

giving

$$\Phi = \frac{ar^n}{n} \cos n\varphi, \qquad \Psi = \frac{ar^n}{n} \sin n\varphi.$$

The streamlines are represented by the curves

$$\Psi = \frac{ar^n}{n} \sin n\varphi = \text{const.}$$

For each value of the constant this is a relation between r and φ. For instance, for $\Psi = 0$ we have $\sin n\varphi = 0$ (excluding the point $r = 0$). Hence $\varphi = k\pi/n$, where $k = 0, 1, 2, \cdots$, and the straight lines through the origin given by $\varphi = \text{constant} = (k\pi)/n$ are streamlines.

We obtain different shapes of streamlines according to the value given to n. The ambiguity arising from the fact that the functions sometimes are many valued can be avoided by replacing some of the straight streamlines by solid walls.

When $n > 2$, we get a pattern similar to that in Fig. 93a or 93b.

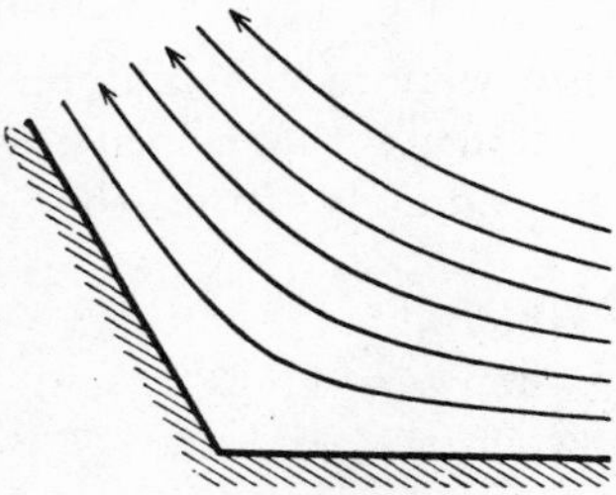

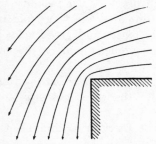

Fig. 94a.

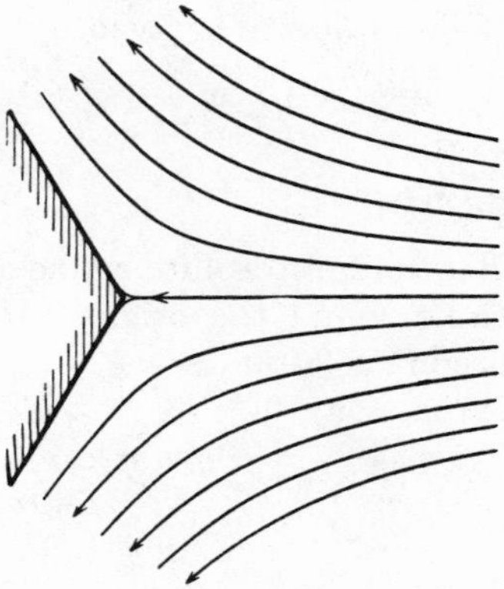

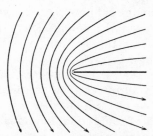

Fig. 94b.

Figs. 94a–b.—Two-dimensional flow: Stream function $F(z) = \dfrac{a}{\frac{3}{2}} z^{\frac{3}{2}}$

When $n = 2$, the form is shown in Fig. 72.
When $1 < n < 2$, refer to Figs. 94a and 94b.
When $n = 1$, the motion is along the real axis (Fig. 90).

Fig. 95.—Two-dimensional flow: Stream function

$$F(z) = \frac{a}{\frac{2}{3}} z^{\frac{2}{3}}$$

Fig. 96.—Two-dimensional flow: Stream function

$$F(z) = \frac{a}{\frac{1}{2}} z^{\frac{1}{2}}$$

When $n = \frac{2}{3}$, the stream moves round a rectangular edge projecting into the fluid, as in Fig. 95.
When $n = \frac{1}{2}$, the stream flows round the edge of a plate as in Fig. 96.

The family of curves Φ = constant, orthogonal to Ψ = constant, is obtained by turning the streamline pattern through an angle $\pi/2n$. These few examples show how motions of outstanding interest can be obtained from really simple functions.

The velocity is given by

$$\overline{\mathbf{w}} = F'(z) = az^{n-1} = ar^{n-1}e^{i(n-1)\varphi},$$

so that

$$|\overline{\mathbf{w}}| = ar^{n-1}.$$

Hence the absolute value of the velocity is constant on each circle round the origin. If we investigate the velocity at the origin, *i.e.*, at $z = 0$, we find immediately from the above expression,

$$\begin{aligned}
\mathbf{w} &= 0, &\quad &\text{when } n > 1; \\
\mathbf{w} &= \text{finite}, &\quad &\text{when } n = 1; \\
\mathbf{w} &= \text{infinite}, &\quad &\text{when } n < 1.
\end{aligned}$$

The fact that when $n < 1$, *i.e.*, when dealing with motion round projecting edges, the velocity at the sharp edge is infinite, is of extreme importance in the practical application of these potential functions. In actual fact the velocities are prevented from becoming infinite at the edge by the action of viscosity, which here again is of utmost importance in a very small part of the fluid where the velocity gradient is large. A vortex is formed at the projecting edge which prevents the velocity from becoming infinite. At the first instant after starting from rest, however, the potential flow actually exists and great velocities occur at the sharp edge. These phenomena will be discussed again in Art. 93.

In the further investigation of the function $F = az^n n$ we shall limit ourselves to a consideration of the three cases $n = -1$, lim $n \to 0$, and lim $n \to \infty$.

78. The Motion Round a Straight Circular Cylinder.—Consider the function

$$F = \frac{a}{z} = \frac{a}{r}(\cos \varphi - i \sin \varphi) = \frac{a}{r}e^{-i\varphi}.$$

The streamlines are

$$\Psi = \frac{a}{r} \sin \varphi = \text{const.}$$

or, making the constant equal to a/C,

$$r = C \sin \varphi.$$

This formula represents a family of circles of diameter C tangent to the real axis at the origin as shown in Fig. 97. A special case is the real axis which is the circle for $C = \infty$.

This type of streamline may be considered as that due to a two-dimensional doublet along the z-axis. The parallel problem of the rotationally symmetrical flow round a sphere has already been investigated by other means in Art. 70 (Fig. 79).

The velocity is given by

$$\overline{\mathbf{w}} = -\frac{a}{z^2} = -\frac{a}{r^2}e^{-2i\varphi}$$

and therefore

$$|\overline{\mathbf{w}}| = \frac{a}{r^2}.$$

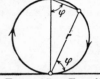

Fig. 97.—Two-dimensional flow: Stream function $F(z) = a/z$ (two-dimensional dipole).

The velocity at the origin therefore becomes infinity to the second order.

Superposing a parallel flow, $F = az$ upon this doublet flow, we have

$$F = a\left(z + \frac{1}{z}\right) = \underbrace{a \cos \varphi\left(r + \frac{1}{r}\right)}_{\Phi} + \underbrace{i\, a \sin \varphi\left(r - \frac{1}{r}\right)}_{\Psi}.$$

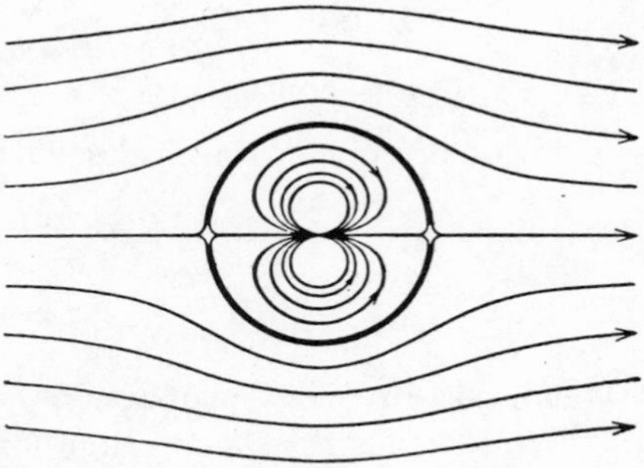

Fig. 98.—Flow round a circular cylinder.

We shall first investigate the streamlines $\Psi = $ constant. If Ψ is zero, we have either $\sin \varphi = 0$ or $r - \frac{1}{r} = 0$. In the first case, $\varphi = 0$ or π, which means that both halves of the real axis are streamlines. In the second case $r = 1$, i.e., the circle of unit radius is also a streamline. The combination of straight line and circle should be considered as a curve of the third

degree. Giving Ψ different constant values, we can obtain φ for any value of r or *vice versa*, and the streamlines can be found point by point, as shown in Fig. 98.

Since only the motion exterior to the unit circle is of physical importance, the above expression gives the potential and the stream functions for the motion round a circular cylinder of unit radius. If the radius is R, the corresponding stream function is

$$F = a\left(z + \frac{R^2}{z}\right).$$

Let us now form the limiting value of the function $\frac{a}{n}z^n$ as n approaches zero, *i.e.*

$$\lim_{n=0} \frac{a}{n}z^n.$$

We have, neglecting an infinite additive constant,

$$F = a \log z = a \log (re^{i\varphi}) = \underbrace{a \log r}_{\Phi} + \underbrace{ia\varphi}_{\Psi}.$$

The streamlines $\Psi = $ constant, or $a\varphi = $ constant, form a pencil of lines through the origin. The curves of constant potential are

$$\Phi = a \log r = \text{const.}$$

or

$$r = \text{const.},$$

and these form a series of concentric circles with the origin as center.

The velocity is

$$\overline{\mathbf{w}} = \frac{a}{z},$$

i.e., the velocity decreases inversely proportional to the distance from the origin. $F = a \log z$ is the stream function for a uniform source along the z-axis, where the quantity of liquid produced at the source per unit time is $2\pi a$ per unit length of the source. The expression for a sink is of course $F = -a \log z$.

Consider the function obtained by multiplying the stream function for a source by the imaginary quantity

$$iF = ia \log z$$

so that the streamlines and potential lines interchange their meaning in this problem. The streamlines are now a set of

circles whose center is the origin, and the equipotentials are straight lines through the origin. The velocity is here

$$\overline{\mathbf{w}} = F'(z) = i\frac{a}{z}$$

or

$$|\overline{\mathbf{w}}| = w = \frac{a}{r}.$$

The line integral $\oint \mathbf{w} \circ d\mathbf{r}$ round one of these circles (streamlines) is equal to $2\pi r w = 2\pi a = $ constant. We therefore realize that the stream function $F = ia \log z$ represents a rectilinear vortex along the axis of z.

Let us now obtain the limiting value of a function involving

$$\frac{a}{n}z^n$$

as n tends to infinity, or more accurately

$$\lim_{n=\infty} a\left(1 + \alpha\frac{z}{n}\right)^n.$$

This is

$$F = ae^{\alpha z} = ae^{\alpha x} \cdot e^{i\alpha y} = ae^{\alpha x}(\cos \alpha y + i \sin \alpha y).$$

Here we have trigonometric functions whose arguments are the y-coordinate and not the angle φ as before.

The first of the streamlines to be considered is $\Psi = 0$, which is

$$\alpha y = k\pi \qquad (k = 0, 1, 2, \cdots),$$

a series of lines parallel to the real axis at a distance π/α from each other. For the domain between two of these lines we have for $\Psi = $ constant,

$$e^{-\alpha x} = \frac{a \sin \alpha y}{\text{const.}}$$

or

$$x = -\frac{1}{\alpha} \ln (\sin \alpha y) + \text{const.}$$

The number pairs (x, y) derived from this equation for a series of values of the constant give the streamlines. The above function—when turned through 90 deg in the positive direction, *i.e.*, the function $F = ae^{i\alpha z}$—can be considered as the potential of a wave motion of small amplitude in two dimensions.

79. The Fundamentals of Conformal Transformation.—In Art. 78 we investigated the motions defined by several simple

analytic functions of the complex variable. We shall now proceed to the converse of this method, by which it is possible to determine the stream function for the motion round a given body.

In order to do this we shall first explain the meaning of conformal transformation from the point of view of function theory. If we start with some analytic function $F(z)$, whose real part is $\Phi(x, y)$ and whose imaginary part is $\Psi(x, y)$

$$F = \Phi + i\Psi,$$

then to every point z, or to every number pair (x, y), there belongs a value of F, *i.e.*, a value of Φ and Ψ. For example, let us take the function discussed in Art. 77,

$$F = \frac{a}{2}z^2 = \frac{a}{2}(x^2 - y^2) + iaxy = \Phi + i\Psi,$$

and, if we consider the (x,y)-plane in Fig. 99 to be thickly crowded

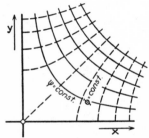

FIG. 99.—The curves $\Phi =$ constant, $\psi =$ constant of the function

$$F = \frac{a}{2}z^2 = \frac{a}{2}(x^2 - y^2) +$$
$$iaxy = \Phi + i\Psi.$$

with Φ and Ψ curves, we realize that every point in the plane can be considered as the intersection of some two curves $\Phi =$ constant, $\Psi =$ constant.

Let us now consider a similar Φ, Ψ-plane, whose rectangular coordinates are Φ and Ψ, then the network of straight lines $\Phi =$ constant and $\Psi =$ constant of this Φ, Ψ-plane represents the $\Phi =$ constant and $\Psi =$ constant network of the x, y-plane. This is called a transformation from one net to the other. Since the sides of the square meshes can be made to converge to zero we can also call the process a transformation from the x, y-plane to the Φ, Ψ-plane, and *vice versa*.

The special case with which we are dealing, where, on account of the validity of the Cauchy-Riemann equations, the mesh of *squares* of the Φ, Ψ-plane transforms into another *square* mesh in the x, y-plane, is called a "conformal transformation." By a conformal transformation we mean a transformation from one plane to another of such a nature that angles in one plane are transformed into equal angles in the same sense in the other plane, and that the ratio of any two contiguous small lengths in

one plane is equal to the ratio of the corresponding lengths in the other plane.   We say that a conformal transformation builds up a second plane which is similar in its infinitesimal elements to the first plane, though corresponding finite areas are not similar, As can be seen from the Cauchy-Reimann differential equations. the transformation which can be effected by any analytic function of the complex variable is conformal over the whole region where the derivative of the function is neither zero nor infinite. When the derivative at some point is either zero or infinite, the corresponding angles round this point are not equal.

In Fig. 99 there is a number pair Φ,Ψ corresponding to every number pair x,y; similarly in Fig. 100, a number pair x,y can be found corresponding to every Φ,Ψ.   In order to show this we take in Fig. 99 a series of different values of x (with y constant), read off the values of Φ and Ψ and insert these at the corresponding points in the Φ,Ψ-plane of Fig. 100.   By repeating this process for a number of y-values we get a family of curves $y = $ constant in the Φ,Ψ-plane.   The corre-

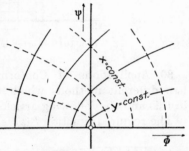

Fig. 100.—The curves $x = $ constant, $y = $ constant, of the function of Fig. 99, when Φ, Ψ are the rectangular coordinates.

sponding curves for $x = $ constant are obtained similarly by reversing the x with the y.

At first we considered the number pair Φ,Ψ as a function of x,y, *i.e.*, F as a function of z (for instance $F = az^2$); now we have conversely z as a function of $F (z = \pm \sqrt{F/a})$.   Since the inverse function is also an analytic function (excluding singular points) the transformation is conformal, and the sets of curves $x = $ constant, $y = $ constant of the Φ,Ψ-plane in Fig. 100 must form a square network.

It may occur however that the expression corresponding to one of the planes is many valued, as in the case of $F = az^2$. In this case, one point of the Φ,Ψ-plane corresponds to two points in the x,y-plane, namely, $z = \pm \sqrt{F/a}$, and the points in the upper half of the x,y-plane fill the whole of the Φ,Ψ-plane, so that the whole x,y-plane covers the Φ,Ψ-plane twice.   In

order to avoid this multiplicity we may, as Riemann has shown, consider the lower half of the x,y-plane to be mapped on a second sheet of the Φ,Ψ-plane which is connected with the first sheet along the positive real axis.

The method of conformal transformation can be applied with success to a great variety of plane motions.

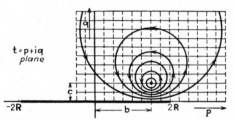

FIG. 101.—Vortex in the $t = p + iq$ plane.

80. Applications of Conformal Transformation.—There are two different methods which we shall investigate rather fully in this and the following articles.

The required function $F(z)$ is given as a function of a function

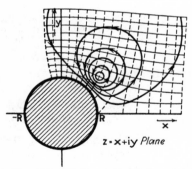

FIG. 102.—Deformation of the vortex of Fig. 101 when the t-plane is transformed into a z-plane so that the piece from $-2R$ to $+2R$ is transformed into a circle of radius R.

$t(z)$, or, which is the same thing, $F(z)$ is given in parametric form:

$$F = F(t), \qquad z = \varphi(t).$$

As an example of this method we can consider the motion of two vortices of equal strength and opposite sign behind a circular cylinder of radius R. (For simplicity only the upper vortex is shown in Fig. 102, the lower one is placed symmetrically with respect to the x-axis.) We consider the z-plane transformed conformally in such a way that the circular contour of the cylinder corresponds to a double straight line (a slit of width 0 and length $4R$), and the exterior of the circle fills the whole of the new plane which we shall call the t-plane, where $t = p + iq$. The interior of the circle is thereby transformed to a second sheet of the Riemann surface.

The streamline pattern of the vortex pair in the t-plane has the form shown in Fig. 101. By making use of the notation of Fig. 101 and the statements made in Art. 78, the stream function $F(t)$ corresponding to this case is

$$F(t) = ia[\ln(t - b + ic) - \ln(t - b - ic)]$$

(one vortex is negative and the other positive, hence the minus sign before one of the logarithms).

As will be shown later, the transformation function from the t-plane to the z-plane (transforming the straight line from $-2R$ to $2R$ in the t-plane into the circle of radius R and center at the origin in the z-plane) is:

$$t = z + \frac{R^2}{z}$$

or

$$z = \frac{t}{2} \pm \frac{1}{2}\sqrt{t^2 - 4R^2}.$$

On the one hand we have $F = F(t)$, and on the other we have $t = t(z)$, so that $F = F[t(z)]$ is given as a function of a function. The required function $F(z)$ is therefore obtained by eliminating t.

In principle the state of affairs is exactly the same as when $F(z)$ is given in the parametric form

$$F = F(t) \qquad \text{and} \qquad z = \varphi(t).$$

If we proceed from the well-known function $F(t)$ representing the motion round a circular cylinder, whose section for simplicity will be taken as that of a unit circle, then, by means of the function

$$z = t + \frac{1}{t},$$

the circle in the t-plane transforms into a straight line on the real axis of the z-plane and the motion round the circle transforms into a flow along the real axis.

The introduction of polar coordinates into the transformation brings these points out quite clearly. If $t = p + iq = re^{i\varphi}$, we get

$$z = re^{i\varphi} + \frac{1}{r}e^{-i\varphi},$$

and along the circle of unit radius this becomes

$$z = e^{i\varphi} + e^{-i\varphi}$$

or

$$z = 2\cos\varphi.$$

Hence when $|t| = 1$, z is real and has a maximum value of $+2$ corresponding to the point $+1$ on the t-plane and a minimum value -2 corresponding to the point -1 on the t-plane. The circumference of the circle is therefore transformed into the straight line from -2 to $+2$ taken twice over. The exterior of the circle is transformed into the whole of the z-plane, and

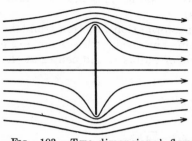

the interior, as has already been said, is mapped on a second sheet of the Riemann surface whose branch points are ± 2. If we had wanted to transform the points ± 1 on the t-plane into the points ± 1 of the z-plane, we could have taken $z = \frac{1}{2}(t + 1/t)$ for our transformation function. This was not done, however, because we want z and t to have the same

FIG. 103.—Two-dimensional flow around a plate at right angles to the stream.

unit of length in the infinite parts of the two planes. If it had been done, we would have found that the unit in the infinite part of the t-plane is twice that in the corresponding part of the z-plane.

In the same way the function

$$z = i\left(t + \frac{1}{t}\right)$$

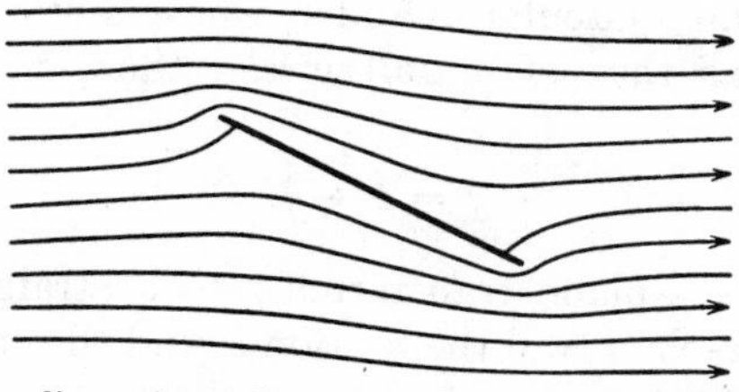

FIG. 104.—Two-dimensional flow around a plate inclined to the stream.

transforms the motion round a circular cylinder in the t-plane into a flow round a plate from $-2i$ to $2i$ along the imaginary axis of the z-plane (Fig. 103). When $|t| = 1$, we get

$$z = (e^{i\varphi} + e^{-i\varphi})i = 2i \cos \varphi.$$

Hence when t lies on the unit circle, z is purely imaginary, the points $t = 1$ (for $\varphi = 0$), $t = +i$ (for $\varphi = \pi/2$), $t = -1$ (for $\varphi = \pi$) and $t = -i$ (for $\varphi = 3\pi/2$) correspond to $z = 2i$, $z = 0$, $z = -2i$, and $z = 0$, respectively.

The linear combination

$$z = a\left(t + \frac{1}{t}\right) + ib\left(t + \frac{1}{t}\right)$$

describes a transformation of a circular cylinder into a flat plate inclined at an angle $\alpha = \tan^{-1}(-b/a)$ to the advancing stream (Fig. 104).

Without entering into great detail, we shall now indicate the way to derive some additional transformations which have been of great importance in airfoil theory.

In the last type of motion considered, namely that of a flat plate inclined to the general stream (Fig. 104), the flow exerts a moment on the plate tending to rotate the plate clockwise. This can be understood from an inspection of the figure if it is remembered that the stagnation points are points of maximum pressure. However, a single resulting force (a lift or drag) cannot be created by any trans-formation of the symmetrical flow round a cylinder of Fig. 98.

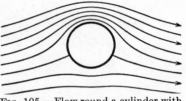

FIG. 105.—Flow round a cylinder with circulation.

If we wish to arrive at a result in which there is such a resultant force, we must either assume a discontinuous motion as will be done in Art. 82 or destroy the symmetry of the motion. This can be done by superposing on the symmetrical flow of Fig. 98 a circulating potential motion as was discussed in Art. 78. By assuming a vortex in the center of the cylinder, we still retain a potential motion by excluding the inside of the cylinder from our consideration. This is justifiable since there is no fluid in the interior of the circle. The streamline pattern thus obtained is shown in Fig. 105. The effect of the vortex is seen as an increase in the velocity above, and a decrease in the velocity below the cylinder, implying a difference of pressure between above and below and correspondingly a lift.

We now follow the method first introduced by Kutta and transform the circle into a line inclined to the general direction of the stream by means of the function given above. In addition we so choose the strength of the vortex at the center of the circle that the rear stagnation point lies on the trailing edge of the plate (Fig. 106). This type of motion produces a

force (that which previously acted on the cylinder) at right angles to the velocity at infinity $\mathbf{w}_\infty$.

FIG. 106.—Combination of parallel flow and circulation round plate inclined to the stream. The circulation is so chosen that the streamline flows smoothly off the trailing edge.

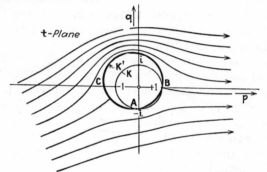

FIG. 107.—Flow round a circle eccentric to the unit circle.

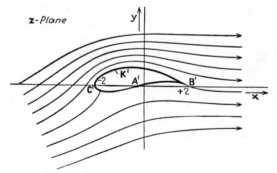

FIG. 108.—Flow round a Joukowsky profile. Conformal transformation from the t-plane of Fig. 107 into a z-plane in which the unit circle of the t-plane is transformed into a straight line from -2 to $+2$.

If we wish to obtain contours very similar to the profiles of modern airfoils, we must use the transformation introduced by Joukowsky. The essentials of this transformation are that we have the unit circle (K), and an eccentric circle (K') tangent to

the unit circle or cutting it in two points A and B (Fig. 107), and that we transform this whole figure in such a way that the unit circle becomes a straight line from -2 to $+2$. Now follow the contour of the circle K' in a counterclockwise direction starting from B, and since it lies outside K and in the upper half of the t-plane, it will transform into a curve in the z-plane which

starts from the point $+2$ and lies in the upper half of the z-plane (Fig. 108). The angle made by the circles K and K' at the point $+1$ of the t-plane is doubled in the z-plane as a result of this transformation. Follow the circumference of K' still further, then the transformed curve must cut the real axis of the z-plane at a point C' corresponding to the point C of the

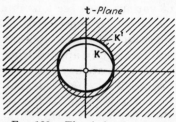

Fig. 109.—The circle whose center is not at the origin transforms into the arc in Fig. 110 by conformal transformation.

intersection of K' with the real axis of the t-plane; the greater the distance of C' to the left of the point -2, the greater must be the corresponding distance of C to the left of -1 in the t-plane. The transformed circle K' must now approach the real axis and from the point A' till $+2$ must lie on the second sheet of the Riemann surface. This corresponds to the fact that K' here

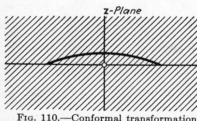

Fig. 110.—Conformal transformation of circle K of Fig. 109 into an arc by the function

$$z = t + 1/t$$

lies in the interior of the unit circle K. The curve obtained by this qualitative transformation has a form which is very similar to the profile of an airfoil, as can be seen from Fig. 108. If in a similar way we transform the motion involving circulation round K in the t-plane into a corresponding motion in the z-plane, and if we choose the strength of vortex so that the fluid flows smoothly off the trailing edge, then we arrive at the type of streamline shown in Fig. 108.

The motion round a convex plate is obtained, as Kutta demonstrated, when the circle K' round which the liquid is flowing has the position shown in Fig. 109. The exterior of such a circle is transformed into the whole of the z-plane, so that the branch

cut is an arc (Fig. 110). The flow round such an eccentric circle is therefore transformed into that round an arc. It is also possible to obtain a transformation in which the arc is inclined to the stream.

81. The Hodograph Method.—We now proceed to another method which enables us to investigate the flow round bodies of various cross sections in a systematic manner. This method may also be considered as a special case of the previous one, since the velocity $\mathbf{w}$ appears as a parameter. This investigation can be usefully applied when it is possible to make some definite statements about the velocity field, which occurs rather often.

Since $\overline{\mathbf{w}} = F'(z)$ is an analytic function of z, the transformation from the $\overline{\mathbf{w}}$-plane to the z-plane is a conformal one. Since on the other hand the transformation from the z-plane to the F-plane is also conformal, the $\overline{\mathbf{w}}$-plane is mapped on the F-plane with equal angles, so that

$$\overline{\mathbf{w}} = \varphi(F)$$

is also an analytic function.

The relation between the stream function and z is usually much too complicated to be written down a priori, but the relation between $\overline{\mathbf{w}}$ and F is often so simple that it is possible to find its analytic expression from a simple physical inspection. If, however, we know the function $\overline{\mathbf{w}} = \varphi(F)$, we can always obtain the desired relation between z and F by integration, because $\overline{\mathbf{w}} = F'(z)$, or $dz = dF/\overline{\mathbf{w}}$. Hence

$$z = \int \frac{dF}{\overline{\mathbf{w}}} + \text{const.}$$

or

$$z = \int \frac{dF}{\varphi(F)} + \text{const.}$$

If the velocity field in the z-plane is known, it is possible to construct a $\overline{\mathbf{w}}$-plane, and this roundabout way through the $\overline{\mathbf{w}}$-plane is often useful for first finding the function $F(\mathbf{w})$, and then, by integration, for finding $F(z)$. In this process we first have to draw the lines $\Psi = \text{constant}$ in the u,v-plane, *i.e.*, we make a diagram in which velocity vectors of all points on some chosen streamline are drawn from a common point like a pencil of rays.

The curve $\Psi = \text{constant}$ in the $\mathbf{w}$-plane is called a "hodograph," a name first introduced by Hamilton, and the method sketched above is called the "hodograph method."

Let us apply this method to some definite example which will make the above general description much clearer. Consider the flow into a so-called "Borda mouthpiece." Without knowing the stream function we can indicate the probable flow of the streamlines from general physical principles (Fig. 111). The fluid flows toward the opening on the left from all sides. In the slit of the mouthpiece itself the velocity has the constant value a provided that the point under consideration is sufficiently far away from the mouth; then $u = a$ and $v = 0$. Outside of it the velocity will decrease with the distance from the mouth, and in the infinitely distant part of the plane, considered as a point A in the function theory, it converges to zero.

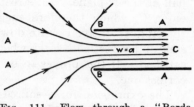

Fig. 111.—Flow through a "Borda mouthpiece."

At the sharp edges of the mouth the velocity becomes infinitely great as we know from Art. 77.

As a result of these observations about the velocity field in the z-plane we shall now attempt to build up the $\mathbf{w}$-plane (Fig. 112). Since all streamlines come from the infinite part of the plane and have zero velocity there, the whole of the infinitely distant part of the z-plane is represented by the zero point in the $\mathbf{w}$-plane. All streamlines in the plane start from this

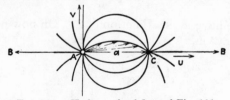

Fig. 112.—Hodograph of flow of Fig. 111.

point. Let us follow the center stream filament of the z-plane, of which, since its direction remains constant, the velocity $\mathbf{w}$ increases from zero to a. In the $\mathbf{w}$-plane, therefore, this stream-line is a straight line from $\mathbf{w} = 0$ to $\mathbf{w} = a$. Let us now follow the streamline coming from infinity from the right along the lower wall of the mouthpiece. The velocities at the points of this streamline keep a constant direction toward B and increase in value as the points approach B. At B the velocity is infinite. After turning the corner its direction is now reversed

and its value falls to a. In the **w**-plane this streamline is the combination of the straight line from A to $-\infty$ and the straight line from $+\infty$ to C. The other streamlines must be approximately of the form shown in Fig. 112, since they must all start from the point $u = 0$, $v = 0$, and end at the point $u = a$, $v = 0$. Thus the upper half of the z-plane corresponds to the lower half of the **w**-plane, and the lower half of the z-plane to the upper half of the **w**-plane. All streamlines emerge from A and finally enter C, which therefore suggests a source and a sink. We therefore tentatively write down the expression for a source and sink and investigate later whether the boundary conditions are satisfied. If this proves to be the case, then the solution is the correct one.

The expression for a source at $w = 0$, and a sink at $w = a$, is derivable from Art. 78, and if c is some constant of the dimensions of a length, we have

$$F(w) = c[\ln w - \ln (w - a)]$$

or

$$e^{\frac{F}{c}} = \frac{w}{w - a},$$

so that

$$w = \frac{a}{1 - e^{-\frac{F}{c}}}.$$

We therefore have w as a function of F. On now performing the second integration,

$$z = \int \frac{dF}{w} + \text{const.},$$

we have

$$z = \int \frac{dF}{a}\left(1 - e^{-\frac{F}{c}}\right) + \text{const.}$$

giving

$$z = \frac{F}{a} + \frac{c}{a}e^{-\frac{F}{c}} + \text{const.}$$

This is the required relation between z and F, which after splitting up into real and imaginary parts appears as

$$z = x + iy = \frac{1}{a}\left[\Phi + i\Psi + ce^{-\frac{\Phi}{c}}\left(\cos \frac{\Psi}{c} - i \sin \frac{\Psi}{c}\right)\right] + \text{const.}$$

Since the constant in this expression simply means a change of origin in the Φ,Ψ-plane, we can put it equal to zero, thus making the real axis coincide with the lower wall of the mouthpiece.

We shall now verify whether this function satisfies the boundary conditions, *i.e.*, whether or not the walls of the mouthpiece are streamlines of the system. We find that $\Psi = 0$ means $ay = 0$, giving the lower wall of the opening, and $\Psi = 2\pi c$ is equivalent to $ay = 2\pi c = \text{constant}$. If the distance between the walls is $y = d$, then, since the constant c is still at our disposal, we can choose it to satisfy the equation $ay = 2\pi c = ad$ or $c = ad/2\pi$. We then see that the upper wall of the Borda mouthpiece is a streamline, so that the function found above indeed satisfies the boundary conditions.

The equation for the center stream filament, the one distant $\pi c/a$ from the real axis, is $\Psi = \pi c$, which is equivalent to

$$ay = \pi c = \text{constant.}$$

In the particular case of the streamline $\Psi = \pi c/2$, y can be expressed explicitly as a function of x. For this value we have

$$ay = \frac{\pi c}{2} - ce^{-\frac{\Phi}{c}},$$

where $\Phi = ax$, and so we get

$$y = \frac{\pi c}{2a} - \frac{c}{a}e^{-\frac{ax}{c}}.$$

Let us now consider the value of the potential function Φ for several values of Ψ. When $\Psi = 0$, *i.e.*, for the streamline corresponding to the lower wall,

$$ax = \Phi + ce^{-\frac{\Phi}{c}}.$$

Figure 113 shows Φ as a function of ax as expressed by the above relation. We shall explain the general character of this curve by following the path of a fluid particle along the lower wall of the opening, corresponding to $\Psi = 0$. A particle on the outer surface of the lower wall at a great distance from the mouth has a small velocity, *i.e.*, a small $\partial\Phi/\partial x$. Hence when x is large, the Φ-curve is nearly parallel to the ax-axis. The velocity or $\partial\Phi/\partial x$ decreases inversely proportional to the distance from the mouth or $\partial\Phi/\partial x = \text{constant}/x$. An integration of this relation shows that Φ becomes logarithmically infinite for large values of x. While the fluid particle approaches the

mouth on the underside of the wall, its velocity increases continuously, which means a continuously increasing slope of the Φ-curve. At the end point of the wall, where the particle turns the corner, the velocity becomes infinite and changes its direction. This means that the tangent to the Φ-curve is

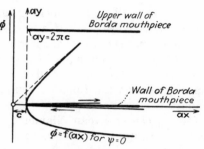

Fig. 113.—The potential function Φ for $\Psi = 0$. (Lower wall of the Borda mouthpiece.)

vertical at the point in which it cuts the ax-axis, so that here $\partial\Phi/\partial x$ is infinite. This point of section with the ax-axis is at a distance c from the origin, because the constant of integration was assumed to be zero. After the particle has moved round the sharp edge ($ax = c$, $y = 0$) into the interior of the mouth-

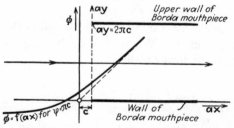

Fig. 114.—The potential function Φ for $\Psi = \pi c$. (Middle streamline of flow in the Borda mouthpiece.)

piece, its velocity is in the positive direction and decreases to a constant limiting positive value a, for, when Φ is positive,

$$\lim_{x = \infty} \frac{\partial\Phi}{\partial x} = a.$$

Hence Φ as a function of ax approaches asymptotically to a line inclined at an angle of 45 deg to the axis of ax.

In following the change of potential on the center stream-line we have to form the expression for ax when $\Psi = \pi c$, namely,

$$ax = \Phi - ce^{-\frac{\Phi}{c}}.$$

Figure 114 gives the shape of this curve. For large negative values of ax, Φ becomes logarithmically infinite, which again is equivalent to a decrease of velocity inversely proportional to the distance from the mouth. The velocity increases more and more on approaching the mouth, till in the slit itself it approaches the value a as a limit. Here again the curve has an asymptote inclined at 45 deg to the axis.

82. Discontinuous Fluid Motions.—Another group of problems to which the method of conformal transformation has been applied with great success deals with so-called "discontinuous" fluid motions. This branch of development was originated in a now classical paper by Helmholtz.[1]

We have an example of discontinuous motion if water flows out of a large tank through a sharp-edged slit as shown in Fig. 115. If the velocities in the outflowing jet are sufficiently large, the effect of gravity can be neglected.

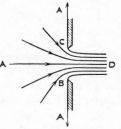

From physical considerations some general statements can be made regarding the

FIG. 115.—Flow through sharp-edged fissure.

shape of the streamlines and about the velocities. In the tank itself, at a great distance from the slit, the flow will be similar to that produced by a sink at the opening. This means that these streamlines approach the slit radially and that the value of the velocity at great distances is inversely proportional to the distance from the slit. The inside of the wall itself is a streamline.

The liquid does not flow completely around the sharp edge of the slit; hence the velocity does not become infinite there. The fluid tears itself away from this point and forms a jet which can be assumed to be horizontal when the velocities are very great. The pressure along the free surface of the jet must therefore be constant and equal to that of the surrounding air. Hence from the Bernoulli equation, $\rho/2w^2$ must also be constant,

[1] HELMHOLTZ, H., On Discontinuous Fluid Motions (German), *Monatsber. Akad. Wiss., Berlin*, p. 215, 1868; or Two Hydrodynamic Essays (German), *Ostwalds Klassiker*, No. 79.

or, in other words, the velocity w on the free surface of the jet must have a constant value.

The source corresponding to the sink is in the infinite part of the plane and is considered as a point A from the point of view of the function theory. The lower edge of the slit is the point B, the upper one C and the infinite part of the jet where the velocity is constant is D.

As a result of this general picture of the velocity field let us now construct the **w**-plane or hodograph. Here we have a

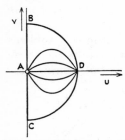

FIG. 116.—Hodograph of flow of Fig. 115.

source A at the origin and a sink D at a distance a along the real axis (Fig. 116) because all streamlines start with velocity $u = 0$, $v = 0$ and end with velocity $u = a$, $v = 0$. Since, as we have seen, the velocity on the free surface is constant, the corresponding curve in the **w**-plane is an arc of a circle of radius a and center at the origin, and since the velocity at B is at angle of $+90$ deg to the direction of the jet, the point B is immediately above A, and similarly, since the direction at C is inclined at an angle -90 deg, the point C is immediately below A, in the **w**-plane. The other streamlines fill up the area of the semicircle as shown in Fig. 116. All streamlines emanate from A and end in D.

We have now to obtain an expression for the function $F(\mathbf{w})$. This is best done by transforming the **w**-plane into another plane in which the semicircle is represented by a straight line. For this purpose we use the function

$$\ln \mathbf{w} = \ln w + i\varphi$$

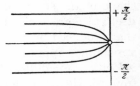

FIG. 117.—Conformal transformation of the half circle of Fig. 115 into an infinite strip.

which transforms the interior of the semicircle into an infinite strip (Fig. 117). By the Schwartz-Christoffel theorem[1] this strip can be transformed into a half plane. Although this method, which depends ultimately on an integration, may appear simple, the application of the Schwartz-Christoffel method is often very difficult on account of the evaluation of

[1] The Schwartz-Christoffel theorem states that every area bounded by straight lines can be transformed into a half plane and that the determination of the transformation function depends on the evaluation of a definite integral.

this integral. The calculation of which the details will not be given here was carried out by Kirchhoff. A result of practical importance obtained by him in this manner is the value of the coefficient of contraction for which he found $\alpha = \pi/(\pi + 2) = 0.61$. This value is in good agreement with experimental observation.

Though it was assumed that the liquid flows out of the slit into a vacuum or gas, the solution is also valid when the jet issues into water which is at rest. This was noted by Helmholtz. The pressure on the surface of the jet is constant and equal to that of the surrounding fluid. We must however impose the limitation that we are dealing with steady motion.

Regarding the occurrence of jets or other discontinuities in real fluids it has been observed that during the first instant after starting from rest the fluid actually does flow around the sharp corners and forms the streamline pattern of ordinary potential motion.

Since the differential equation $\Delta\Phi = 0$ of this motion is also the equation of the electric potential, we can easily determine the form of the streamlines experimentally (Fig. 118).

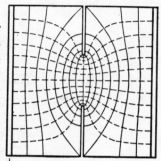

Fig. 118.—Curves of constant electric potential and the family of orthogonal trajectories (broken lines).

This is done by applying an electric potential to the edges of a metal plate at a sufficiently great distance from the deep double slit in the plate. It is very important that the metal rods at the sides of the plate, where the potential is applied, be good conductors so that the potential is constant along the edge. The curves of equal potential can then be obtained by testing the metal surface with some suitable instrument. The curves of the orthogonal trajectories which can then be drawn give the streamline pattern.

But, as we have said, this distribution of velocity occurs only at the first instant of motion. Immediately after this the boundary layer grows, as we shall see in detail in Art. 92, and a surface of discontinuity across which there is a jump of velocity spreads from the sharp edges.

Since this surface of discontinuity is unstable, as will be shown in Art. 93, it breaks up into vortices which immediately

influence the type of motion and disarrange it.   Water jets in air, however, retain their characteristics for a much longer period on account of the smaller density of air and on account of the surface-tension in the water.

Helmholtz called attention to the fact that surfaces of discontinuity are formed when liquids flow round solid bodies.   In connection with this Kirchhoff then developed a general method of investigating the motion round bodies bounded by straight lines.

Among other things he worked out in detail the case of a flat plate at right angles to an advancing stream (Fig. 119).   Calling

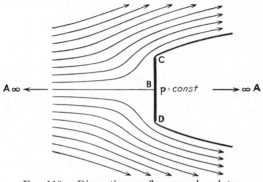

Fig. 119.—Discontinuous flow round a plate.

the infinite part of the plane A, and assuming that there $w = a$, we perceive that the stream approaches the plate from the left in the usual potential type of flow, having zero velocity at the stagnation point B.   Then it detaches itself from the plate in two surfaces of discontinuity at C and D.   Behind the plate the pressure is that of the undisturbed fluid.   We are obliged to make this last assumption about the pressure, because with a pressure smaller than that the two branches C and D of the surface of discontinuity would intersect behind the body.   The undisturbed fluid is called the "dead-water."   Since constant pressure in the dead-water region implies constant pressure over the surface of discontinuity, the Bernoulli equation tells us that the velocity must be constant over this surface.   Also since the surface stretches out into the infinite part of the plane, its velocity must be equal to the velocity a at infinity.   Hence the central streamline approaches the plate symmetrically starting with velocity $w = a$ and decreasing to zero at B.   At

this point the streamline branches off to C and D, the velocity increases, and at C and D attains the value $w = a$. From this point it remains constant in magnitude, but its direction changes gradually and finally it becomes parallel to the general direction of the stream.

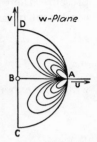

This symmetrical streamline picture in the z-plane is now transformed to the hodograph in the w-plane (Fig. 120). We go from A (at a distance a from the origin) to B, then to D and C along the positive and negative branches of the imaginary axis, and then back to A along the semicircle of radius a. The paths of some of the other streamlines are also shown in Fig. 120. They all must start from A and return to A; in the z-plane of Fig. 119 they all start at infinity

FIG. 120.—Hodograph of flow of Fig. 119.

(to the left) and proceed again to infinity (to the right). A small loop of the closed curve in the hodograph means that the velocity along the streamline departs but little from the value a and from the horizontal direction; thus a small looped hodograph corresponds to a streamline far from the plate in Fig. 119.

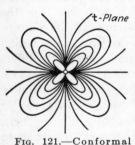

FIG. 121.—Conformal transformation of the w-plane of Fig. 120 into a t-plane in such a way that the interior of the half circle fills half of the t-plane. The semicircle transforms into the real axis of the t-plane.

The w-plane can now be transformed on to a t-plane such that the boundary of the semicircular area becomes the real axis (Fig. 121). This function in the t-plane is the inverse of the function used in Fig. 92; the zero of one function corresponds to infinity in the other and *vice versa*. We therefore have

$$F = \frac{C}{t^2}.$$

Without going into further details we just mention that the treatment of discontinuous fluid motions has been greatly developed in recent years, particularly by Levi-Civita, Cisotti, and Villat.

It is of extreme importance to note that Kirchhoff's investigation of the discontinuous motion past a plate gives a definite value for the resistance. This is an important step ahead, since with a continuous potential motion we always obtain **zero resistance**.

If A be the area of a section of the plate of length l (perpendicular to the plane of the paper), D the resisting force or "drag" experienced by this section of the plate, and $\rho w^2/2$ the stagnation pressure, then Kirchhoff's calculation gives for the dimensionless coefficient of resistance,

$$\frac{D}{A\rho\frac{w^2}{2}} = \frac{2\pi}{\pi + 4} = 0.880,$$

while the experimental value is

$$\frac{D}{A\rho\frac{w^2}{2}} = 2.0.$$

The large discrepancy between these two values is due to the

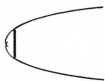

Fig. 122.—Two arcs of parabolas of slightly differing foci as an approximation to the surface of separation of the Kirchhoff flow.

fact that in actual motion the surface of discontinuity is unstable and quickly breaks up into separate vortices. Hence the Kirchhoff type of motion cannot persist. In the actual case there is a great reduction in pressure behind the plate which accounts for the increased resistance. Because of the dissolution of the surface of discontinuity the actual streamline pattern behind the plate differs entirely from the theoretical one of Fig. 119. Theoretically the surface of discontinuity extends to infinity somewhat in the style of two parabolic arcs (or more accurately, two arcs of parabolas with slightly differing foci), as in Fig. 122, but in the actual experiment the surface seems to close up somewhat behind the plate and mingle with the generally irregular vortices. On account of the internal friction of the fluid the irregularities in velocity die out, so that far behind the plate we again have an approximately undisturbed fluid motion.

CHAPTER XII

VORTEX MOTION

83. The Kinematics of Vortex Motion.—In contrast to the type of flow discussed previously, this chapter deals primarily with motions in which the rotation differs from zero at every point or in certain parts of the fluid region:

$$\text{rot } \mathbf{w} \neq 0.$$

Stokes's theorem (Art. 45) states that

$$\iint^{\mathbf{A}} \text{rot } \mathbf{w} \circ d\mathbf{A} = \oint^{\mathbf{C}} \mathbf{w} \circ d\mathbf{r},$$

where **C** is the boundary curve of the surface **A**. In general, therefore, the line integral along a closed curve in rotational motion is not zero. In Art. 72, dealing with plane potential motion with circulation, it was seen that the line integral along a closed curve only differs from zero when the curve surrounds a singular point. This constitutes the fundamental difference between potential motion involving circulation and rotational motion, as explained before in Art. 72.

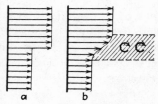

FIGS. 123a–b.—The surface of separation of a nonviscous fluid (a) changes to a layer with rotation on assuming a finite internal friction (b).

So far we have discussed two different ways in which vortices can be generated in a fluid of small viscosity. First, as explained in Art. 55, when liquid flows round blunt-nosed bodies, the substance of the boundary layer, which is always in a state of rotational motion, penetrates into the interior of the fluid. Secondly, when the fluid flows around a sharp edge, a surface of separation is formed which shows a discontinuity in velocity in either value or direction. This type of discontinuous velocity profile is shown in Fig. 123a, but on account of internal friction, however small, the velocity profile changes to that shown in Fig. 123b, which contains a liquid layer with rotation. In

189

Art. 93 we shall return to a discussion of the properties of these surfaces of discontinuity and the reason for their formation.

It was seen in Art. 72 that rot **w** is a measure of the rotation of a fluid element. It is zero everywhere in the case of potential motion, except possibly in a few distinct singular points. Let **q** denote the vector of rotation of a fluid element and therefore q its angular velocity, then we have from Art. 42,

$$\text{rot } \mathbf{w} = 2\mathbf{q}.$$

Many simplifications can be introduced by the use of the vector of rotation **q** instead of the velocity vector **w**. It is important to note that

$$\text{div } 2\mathbf{q} = \text{div rot } \mathbf{w} = \nabla \circ \nabla \times \mathbf{w} \equiv 0.$$

This means that the rotation vector has the geometrical properties of the velocity field of an incompressible fluid. All kinematic theorems dealing with incompressible fluids can thus be applied to fields of the rotation vector. The streamlines of an incompressible fluid therefore correspond to lines of rotation, or vortex lines, of which the direction at every point is that of **q**, *i.e.*, that of the axis of rotation. In the same way that streamlines cannot end abruptly in the interior of the fluid, rotation lines or vortex lines also cannot end abruptly. They must either be closed curves, or they must end on the boundaries or free surfaces.

Stokes's theorem is

$$\int\int^{\mathbf{A}} \text{rot } \mathbf{w} \circ d\mathbf{A} = \int\int^{\mathbf{A}} 2\mathbf{q} \circ d\mathbf{A} = \oint^{\mathbf{C}} \mathbf{w} \circ d\mathbf{r} = \Gamma.$$

The physical meaning of the surface integral is the flow of the rotation vector through the surface **A**; this is called the "intensity of vorticity." It is equal to the circulation Γ along the boundary.

If it is possible to construct a closed curve or thin tube from the lines of rotation, then this tube is called a "vortex filament." Since div **q** $= 0$, the flow of **q** through the tube (*i.e.*, the intensity of vorticity) is the same at every point of the vortex filament.

The rotation rot **w** can be assumed constant over a sufficiently small area. When the vortex filament is thin, we have as an approximation that $\Gamma = 2\mathbf{q} \circ d\mathbf{A}$. On account of the constancy of the rotation Γ along the length of the vortex filament, the value of the angular velocity **q** is inversely proportional to the cross-sectional area of the vortex filament.

It occasionally happens that the rotation in some filament-like region differs from zero while the rest of the fluid is irrotational. We then speak of a vortex filament surrounded by a potential motion. The intensity or strength of this vortex is the value of the line integral of the velocity taken along a closed curve surrounding the vortex. The closed curve can be taken anywhere in the fluid as long as it surrounds the vortex, for the value of the integral is constant.

84. Thomson's Theorem on the Permanence of Circulation.— The theoretical treatment of vortex motion commenced with the classical paper of Helmholtz[1] "On Integrals of the Hydrodynamic Equations Corresponding to Vortex Motions," 1858. Before discussing the theorems of Helmholtz we shall first derive a theorem found by Sir William Thomson (Lord Kelvin) inspired by the work of Helmholtz.

To this end we form the line integral of the velocity along a closed "fluid line" and investigate the rate of change of the value of this integral with respect to time. Since our path of integration is a "fluid line," *i.e.*, a line always composed of the same fluid particles, we must take the substantial differential coefficient:

$$\frac{D}{dt}\oint^{C} \mathbf{w} \circ d\mathbf{r}.$$

The path of integration is a closed curve, therefore a differentiation with respect to the boundary does not enter so that we can write

$$\oint^{C}\frac{D\mathbf{w}}{dt}\circ d\mathbf{r} + \oint^{C}\mathbf{w}\circ\frac{D(d\mathbf{r})}{dt}.$$

By applying Euler's equation (page 110), the first of these integrals becomes

$$\oint^{C}\frac{D\mathbf{w}}{dt}\circ d\mathbf{r} = \oint^{C}\mathbf{g}\circ b\mathbf{r} - \oint^{C}\frac{\text{grad } p}{\rho}\circ d\mathbf{r}.$$

If we now assume an irrotational field of force, *i.e.*,

$$\mathbf{g} = \text{grad } U,$$

and further that the density is a function of the pressure only, *i.e.*, that the fluid is homogeneous but compressible,

$$\int\frac{dp}{\rho} = P(p),$$

[1] See footnote, p. 183.

then the first indefinite integral reduces to

$$\int \frac{D\mathbf{w}}{dt} \circ d\mathbf{r} = U - P.$$

The integrand of the second integral can be transformed as follows: Since time and space differentiation are independent,

$$\frac{D}{dt}d\mathbf{r} = d\frac{D\mathbf{r}}{dt}.$$

But $D\mathbf{r}/dt$ is the rate of change of position of the fluid element with respect to time, *i.e.*, the velocity $\mathbf{w}$; and therefore

$$\frac{D}{dt}d\mathbf{r} = d\mathbf{w}.$$

This gives

$$\mathbf{w} \circ \frac{D}{dt}d\mathbf{r} = \mathbf{w} \circ d\mathbf{w} = d\frac{\mathbf{w}^2}{2},$$

so that the second integral is $\mathbf{w}^2/2$.

Hence the rate of change of the value of the line integral along a fluid line with respect to time is

$$\frac{D}{dt}\int \mathbf{w} \circ d\mathbf{r} = \frac{\mathbf{w}^2}{2} + U - P.$$

(Incidentally it should be noted that the right-hand side of this equation is not that of the Bernoulli equation, which is $\mathbf{w}^2/2 + P - U = \text{constant}$.)

If now we move from one point A on the fluid line to some other point B on the same line, we have

$$\frac{D}{dt}\int_A^B \mathbf{w} \circ d\mathbf{r} = \left[\frac{\mathbf{w}^2}{2} + U - P\right]_A^B.$$

If, however, we are dealing with a closed curve, B coincides with A, and assuming that $\mathbf{w}$ is continuous (*i.e.*, that there are no surfaces of discontinuity), we get

$$\frac{D}{dt}\oint \mathbf{w} \circ d\mathbf{r} = 0$$

giving

$$\oint \mathbf{w} \circ d\mathbf{r} = \text{const.} = \Gamma.$$

This is the theorem of Sir William Thomson, stating that the line integral along a closed fluid curve in a homogeneous non-

viscous fluid is constant for all time, if the fluid is under the influence of an irrotational field of force.

If we take account of the fact that, when a fluid is at rest, the circulation is zero for every closed curve (the velocity is zero everywhere), it follows that every motion developed in this liquid under the action of an irrotational field of force has zero circulation along these lines. Since we now can consider every point of the liquid at rest as being surrounded by any fluid line we care to choose, and since the absence of circulation along this line is equivalent to a lack of rotation, it follows from Thomson's theorem that all motions developed from rest in this way are irrotational, *i. e.*, they are potential motions.

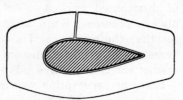

FIG. 124.—A strut-like body surrounded by a closed fluid line which does not contain the body.

The statement that circulation cannot be produced in a motion developed from a homogeneous non-viscous fluid by the action of an irrotational field of force is true only for regions surrounded by fluid lines which were closed curves when the liquid was at rest. Only when this condition is satisfied, can we conclude to a permanent absence of rotation. There are, however, many types of motion developed from rest in which there are surfaces in the liquid that do not lie in the interior

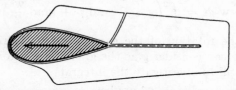

FIG. 125.—The fluid line of Fig. 124 is deformed by the motion. From the sharp trailing edge there projects a surface which does not lie inside the closed curve.

of a region to which Thomson's theorem is applicable. These cases always deal with the confluence of fluid particles that were previously separated. As an example, let us consider the motion round a slender strut-like body with a sharp trailing edge, and let us assume that before the motion commences it is surrounded by a closed curve which itself does not contain the strut (Fig. 124). After the motion has been under way for some time, the closed curve deforms into the shape of Fig. 125, of which the most

important characteristic is a thin region projecting from the trailing edge which does not lie within the fluid line. Thomson's theorem cannot be applied to this thin region, which constitutes the surface of confluence. In such surfaces of confluence of previously separated fluid particles the change of velocity from one side of the surface to the other may be zero as in this example, or there may be a discontinuity. Then the surface is called a "surface of discontinuity," which can be considered as a surface of distributed vorticity (Art. 92). A surface of this type is always produced when an airfoil is in motion. We therefore see that the existence of a surface of discontinuity with transverse or longitudinal discontinuities in velocity does not contradict the classical hydrodynamics of a non-viscous fluid.

Since no assumption was made about compressibility, Thomson's theorem is valid for compressible and incompressible fluids.

85. Extension of Thomson's Theorem to the Case of Non-homogeneous Fluids by V. Bjerkness.—Primarily for application to meteorology, Thomson's theorem has been extended by V. Bjerkness[1] to the case of non-homogeneous fluids and has been given a simple geometrical significance. Since

$$d\frac{\mathbf{w}^2}{2} = \text{grad } \frac{\mathbf{w}^2}{2} \circ d\mathbf{r},$$

we have

$$\frac{D}{dt}\oint^{c} \mathbf{w} \circ d\mathbf{r} = \oint^{c} \left(\mathbf{g} - \frac{1}{\rho} \text{ grad } p + \text{grad } \frac{\mathbf{w}^2}{2} \right) \circ d\mathbf{r}.$$

As long as the rotation of the earth, *i.e.*, the Coriolis force is neglected, we can consider $\mathbf{g}$ as the gradient of a force function.

We have still to investigate the integral

$$\oint^{c} \frac{\text{grad } p}{\rho} \circ d\mathbf{r}$$

of the last equation. Since the atmosphere generally must be considered as a non-homogeneous fluid, *i.e.*, since density is dependent on position in addition to pressure, the surfaces of equal pressure and density $p = $ constant and $\rho = $ constant are not usually identical. Although the integral

$$\oint^{c} \frac{\text{grad } p}{\rho} \circ d\mathbf{r}$$

[1] BJERKNESS, V., "Lectures on Hydrodynamic Forces Acting at a Distance" (German), Leipzig, 1900–1902.

is zero when dealing with homogeneous fluids, it is different from zero when the fluid is non-homogeneous.

Let us transform the line integral into a surface integral by means of Stokes's theorem:

$$\oint^{C} \frac{\operatorname{grad} p}{\rho} \circ d\mathbf{r} = \int\int^{A} \operatorname{rot} \frac{\operatorname{grad} p}{\rho} \circ d\mathbf{A},$$

where the form of the surface **A** is arbitrary so long as it has **C** for its boundary. This agrees with the nature of rotation in that the divergence of a vector field, which itself is the rotation of another vector field **a**, is zero: div rot **a** = 0 (see Art. 47, Case 2).

By making use of the relation

$$\operatorname{rot} \frac{\operatorname{grad} p}{\rho} = \boldsymbol{\nabla} \times \left(\frac{1}{\rho}\boldsymbol{\nabla}p\right) = \boldsymbol{\nabla}\frac{1}{\rho} \times \boldsymbol{\nabla}p + \frac{1}{\rho}\underbrace{\boldsymbol{\nabla} \times \boldsymbol{\nabla}p}_{0}$$

$$= \operatorname{grad} \frac{1}{\rho} \times \operatorname{grad} p$$

and also of the relation

$$\oint^{C}\left(\operatorname{grad} U + \operatorname{grad} \frac{\mathbf{w}^2}{2}\right)\circ d\mathbf{r} = 0,$$

we find for the rate of change of the line integral:

$$\frac{D}{dt}\oint^{C} \mathbf{w}\circ d\mathbf{r} = \int\int^{A}\left(\operatorname{grad} \frac{1}{\rho} \times \operatorname{grad} p\right)\circ d\mathbf{A}.$$

Bjerkness has given a simple geometrical interpretation of this equation. Let us draw equidistant members of the families of surfaces $p = $ constant and $1/\rho = $ constant, and so obtain a series of tubes, the cross section of one of which is shown shaded in Fig. 126. The cross-sectional area a of such a tube is

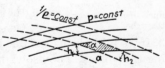

Fig. 126.—Cross section of tube system formed by the surfaces $p = $ constant and $1/\rho = $ constant.

$$a = \frac{h_1}{\sin \alpha}h_2,$$

since h_1 and h_2 are parallel to the directions of grad p and grad $1/\rho$ respectively. Moreover with constant difference between the values of p for two consecutive lines the distance h_1 is inversely

proportional to grad p, *i.e.*, for large grad p the spacing h_1 is small. This gives the relations

$$h_1 = \frac{\text{const.}}{\left|\text{grad }\dfrac{1}{\rho}\right|}, \qquad h_2 = \frac{\text{const.}}{|\text{grad }p|}.$$

This makes the cross section a of a tube equal to

$$a = \frac{\text{const.}}{\left|\text{grad }\dfrac{1}{\rho}\right| \cdot |\text{grad }p| \cdot \sin \alpha} = \frac{\text{const.}}{\left|\text{grad }\dfrac{1}{\rho} \times \text{grad }p\right|}.$$

Making the constant in this result equal to unity merely means that we have made a special choice of the interval between consecutive p or $1/\rho$ surfaces. Then the rate of change of the line integral with respect to time is

$$\frac{D}{dt}\oint^{\mathbf{C}} \mathbf{w} \circ d\mathbf{r} = \int\int^{\mathbf{A}} \frac{d\mathbf{A}}{a}.$$

After performing the integration on the right-hand side, we see that the change of circulation per unit time is equal to the num-

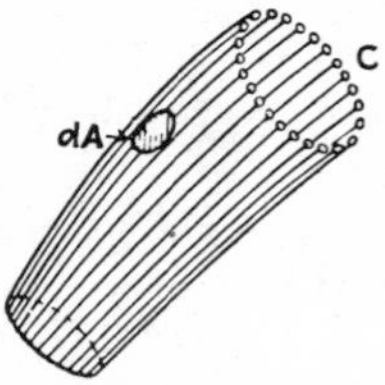

FIG. 127.—Vortex lines forming a vortex tube.

ber of tubes surrounded by the curve **C**.

86. The Dynamics of Vortex Motion.—The dynamical theory of vortex motion can be based on the theorem of Thomson which was derived in Art. 84. The main results that will be found in this article are that a vortex filament is always composed of the same fluid elements and that its strength is constant not only with respect to space (see Art. 83) but also with respect to time.

If we take the closed line integral along any fluid line surrounding the vortex, Thomson's theorem tells us that this line integral (the circulation) is constant with respect to time. We shall now prove that this fluid line always surrounds the vortex filament and is never cut by it.

Consider a group of vortex lines all passing through the closed curve **C** of Fig. 127, where this curve is so chosen that it does not include any singularities. These vortex lines form a vortex tube, of which the interior is a vortex filament. We know also that the flow through such a vortex tube must be constant, since div $\mathbf{q} = 0$, and therefore the vortex strength is constant along the tube at some one instant of time (Art. 83).

A new consideration entering into the argument is that we can apply Thomson's theorem to any arbitrary small element $d\mathbf{A}$ of the wall of the vortex tube (Fig. 127). The flux of the rotation vector across this small element, *i.e.*, the circulation along its contour, is zero since $d\mathbf{A}$ is a part of the wall of the vortex tube which cannot be cut by any line of rotation. But according to Thomson's theorem the circulation, and therefore the flow through this surface, considered as a fluid surface, must always be zero. By piecing together all these areas $d\mathbf{A}$ and so obtaining the whole vortex tube, we show that the total flow through its surface, considered as a fluid surface, must always remain zero. In other words, those fluid elements which at one time form a vortex tube form it for all time, or expressed still shorter: once a vortex always a vortex. It therefore follows that the fluid particles within the vortex tube always remain there. A vortex filament is thus always composed of the same fluid elements. We can of course consider a vortex filament contracted to a single vortex line and so realize that every vortex line is always composed of the same fluid particles.

We know that a fluid line which surrounds a vortex tube at one instant does so for all time. But Thomson's theorem shows that the line integral along a closed curve is constant, and, as shown above, a fluid line enclosing a vortex filament always does so; consequently we have proved the theorem that the strength or vorticity of a vortex filament is always constant and that the vortex filament itself is always composed of the same fluid elements. This theorem was first discovered and proved by Helmholtz in 1858. The above proof is due to Thomson.

87. The Vortex Theorems of Helmholtz.—In his original paper, Helmholtz stated his vortex theorems as follows:

If there exists a potential for all forces acting on a non-viscous fluid, then the following theorems are true:

1. No fluid particle can have a rotation if it did not originally rotate.

2. Fluid particles which at any time are part of a vortex line always belong to that same vortex line.

3. The product of the cross-sectional area and of the angular velocity of an infinitely thin vortex filament is constant over the whole length of the filament and keeps the same value even when the vortex moves. The vortex filaments must therefore be either closed tubes or end on the boundaries of the fluid.

Helmholtz's proof which will now be given assumes a homogeneous and incompressible fluid and thus is less general than Thomson's proof which holds for compressible fluids as well. Helmholtz's proof starts from Euler's equation which contains the relation between pressure and velocity. Since nothing can be said about the pressure, it is eliminated by forming the rotation of Euler's equation:

$$\text{rot} \left(\frac{D\mathbf{w}}{dt} - \text{grad } U + \frac{1}{\rho} \text{ grad } p \right) = 0.$$

Since

$$\frac{D\mathbf{w}}{dt} = \frac{\partial \mathbf{w}}{\partial t} + \mathbf{w} \circ \nabla \mathbf{w},$$

and from page 123,

$$\mathbf{w} \circ \nabla \mathbf{w} = \text{grad } \frac{\mathbf{w}^2}{2} - \mathbf{w} \times \text{rot } \mathbf{w},$$

we have, on writing rot $\mathbf{w} = 2\mathbf{q}$,

$$\frac{D\mathbf{w}}{dt} = \frac{\partial \mathbf{w}}{\partial t} + \text{grad } \frac{\mathbf{w}^2}{2} - \mathbf{w} \times 2\mathbf{q}.$$

Upon assuming that the fluid is homogeneous, *i.e.*,

$$\frac{1}{\rho} \text{ grad } p = \text{grad } P,$$

and by making use of the fact that the rotation of a gradient is zero, this becomes

$$\text{rot } \frac{D\mathbf{w}}{dt} = \text{rot } \frac{\partial \mathbf{w}}{\partial t} + \text{rot } (2\mathbf{q} \times \mathbf{w}) = 0.$$

Or, since

$$\text{rot } \frac{\partial \mathbf{w}}{\partial t} = \frac{\partial}{\partial t} \text{ rot } \mathbf{w} = 2\frac{\partial \mathbf{q}}{\partial t},$$

and since

$$\begin{aligned}
\text{rot } (\mathbf{q} \times \mathbf{w}) &= \nabla \times \mathbf{q} \times \mathbf{w} = \nabla \circ (\mathbf{wq} - \mathbf{qw}) \\
&= (\nabla \circ \mathbf{wq} + \mathbf{w} \circ \nabla \mathbf{q}) - (\nabla \circ \mathbf{qw} + \mathbf{q} \circ \nabla \mathbf{w}) \\
&= \mathbf{q} \text{ div } \mathbf{w} + \mathbf{w} \circ \nabla \mathbf{q} - \mathbf{w} \text{ div } \mathbf{q} - \mathbf{q} \circ \nabla \mathbf{w},
\end{aligned}$$

and since, also,

$$\text{div } \mathbf{q} = \tfrac{1}{2}(\text{div rot } \mathbf{w}) \equiv 0,$$

we get

$$\text{rot } (\mathbf{q} \times \mathbf{w}) = \mathbf{q} \text{ div } \mathbf{w} + \mathbf{w} \circ \nabla \mathbf{q} - \mathbf{q} \circ \nabla \mathbf{w}.$$

Combining this with

$$\frac{\partial \mathbf{q}}{\partial t} + \mathbf{w} \circ \nabla \mathbf{q} = \frac{D\mathbf{q}}{dt},$$

we find the relation

$$\frac{Dq}{dt} = q \circ \nabla w - q \, \text{div } w.$$

This is Helmholtz's starting equation. For further investigation the fluid must be assumed to be incompressible so that div $w = 0$. Herein lies the superiority of Thomson's proof over that of Helmholtz, for Thomson does not postulate incompressibility; he assumes only homogeneity, and the fluid may be compressible. We therefore are left with

$$\frac{Dq}{dt} = q \circ \nabla w.$$

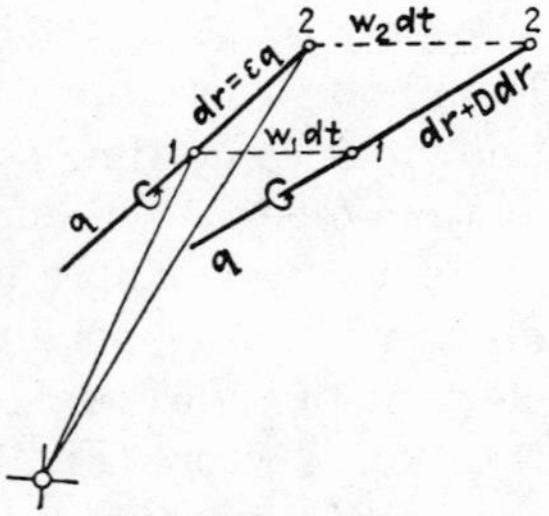

The geometrical meaning of this equation is as follows: Consider two closely neighboring fluid elements 1 and 2 whose joining line has the length ϵq and which lies along the axis of rotation of the fluid particles at the time t_1. It is then obvious that

$$dr = \epsilon q.$$

(See Fig. 128). If the velocity of the first particle is w_1 and that of the second w_2, then, since w is a regular function of r, the first term of the Taylor expansion with respect to r gives

Fig. 128.—If the line joining two closely neighboring fluid elements 1 and 2 lies in the direction of the axis of rotation of these elements at any one instant of time, then the line will remain the direction of the axis of rotation for all time.

$$w_2 - w_1 = dr \circ \nabla w.$$

After a time dt, both fluid particles have changed their position. Particle 1 has a radius vector $r_1 + w_1 dt$, and the corresponding radius vector for particle 2 is $r_2 + w_2 dt$, so that the "substantial" change in dr is

$$D dr = w_2 dt - w_1 dt.$$

Therefore the substantial change per unit time of the vector joining the two particles is

$$\frac{D}{dt} dr = w_2 - w_1,$$

which can be written as

$$\frac{D}{dt} dr = dr \circ \nabla w$$

or as

$$\frac{D}{dt} dr = \epsilon q \circ \nabla w.$$

We know also that the rotation vector has changed during the time dt, and from the Helmholtz equation we get the change per unit time as

$$\frac{D\mathbf{q}}{dt} = \mathbf{q} \circ \nabla \mathbf{w}.$$

Combining this with the equation immediately preceding, we see that

$$\frac{D}{dt} d\mathbf{r} = \epsilon \frac{D}{dt} \mathbf{q}.$$

Helmholtz therefore concluded that if the ratio of $d\mathbf{r}$ to $\mathbf{q}$ was known at some time $t = t_0$, the proportionality would hold good at time $t_0 + dt$, i.e., for all time.

Hence the changes in length and direction of the line joining any two closely neighboring points on a vortex line during an element of time are exactly equal to those of the corresponding rotation vector (Fig. 128). Therefore if any fluid elements are on the vortex line at any time they always remain there. Further, since $d\mathbf{r}$ is proportional to $\mathbf{q}$ and since $d\mathbf{r}$ is inversely proportional to the cross-sectional area da of the vortex tube on account of incompressibility, the product of the area of the tube and the corresponding rotation at the point under consideration must be constant. This constancy of rotation is nothing else but Kepler's law of "equal areas in equal times." This law tells us that in the absence of external rotational forces the angular velocity is inversely proportional to the moment of inertia. Actually, if the deformation of the fluid element is small, the moment of inertia is proportional to the cross-sectional area, assuming that the fluid particle originally was rotationally symmetrical.

The nomenclature at present in use differs from that adopted by Helmholtz. The quantity we call "rotation" was named "vortex" by Helmholtz. This use of the word vortex brings us into conflict with the usual application of the word, which to us implies a circulating fluid motion. According to Helmholtz the pure slipping laminar flow of a viscous fluid would be called a "vortex motion," which definitely contradicts general usage. In this book the word vortex is applied solely to the circulatory motion round an isolated vortex filament (see Art. 71); the laminar motions possess rotation but no vortex.

88. The Velocity Field in the Neighborhood of an Isolated Vortex; the Law of Biot and Savart.—We are now going to investigate the action of an individual vortex on its immediate neighborhood, *i.e.*, we shall consider the velocity field carried along by the vortex in the liquid outside the vortex filament. Since it is assumed that $q = 0$ in the neighborhood of the vortex, the motion is a potential one and we now proceed to find its potential function. The simply continuous space is split into two regions when we assume that the space occupied by vortex filament forms a separate domain.

The simplest case is that in which the vortex is so concentrated that it can be assumed to be linear.

If we cut the fluid along an arbitrary surface which has the vortex line for boundary (Fig. 129), then we make a simply

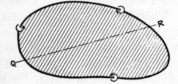

Fig. 129.—Arbitrary surface containing the closed vortex line.

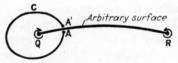

Fig. 130.—Many-valued potential. Whenever the closed path of integration cuts the arbitrary surface there is a discontinuity in the potential.

connected space from one that was previously doubly connected. The line integral of the velocity along any closed path that does not cut the arbitrary surface is zero:

$$\oint^C \mathbf{w} \circ d\mathbf{r} = 0.$$

The integral between two points O and A is

$$\int_O^A \mathbf{w} \circ d\mathbf{r} = \Phi_A - \Phi_O.$$

If the path of integration is that shown in Fig. 130, from A to A', then

$$\int_A^{A'} \mathbf{w} \circ d\mathbf{r} = \Phi_{A'} - \Phi_A.$$

If now we allow A' to converge to A, we get

$$\oint \mathbf{w} \circ d\mathbf{r} = \Gamma,$$

so that on piercing the arbitrary surface by a closed path there is a sudden increase of potential Γ. This discontinuity in the

potential is the same for all points of the surface, as can be seen from Fig. 131. If the path of integration is so chosen that the arbitrary surface is cut at BB' instead of at AA', then, by adding branch paths $A'B'$ and AB, we add to the original path of integration another closed path $AA'B'BA$. This path lies in a simply connected space, for the arbitrary surface can be taken in some

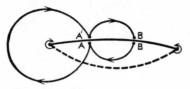

other position, the broken curve for example. Here the integral along the closed path $A'B'BA$ is zero so that $\Gamma_B = \Gamma_A$.

Fig. 131.—Many-valued potential. The change of potential produced in passing through the arbitrary surface is the same at every point.

By continually going through the separating wall of the arbitrary surface, the potential assumes an infinite number of values and increases by Γ with every circuit round the filament. Therefore Γ is seen to be the strength of the vortex.

There is an analogous phenomenon in the magnetic field produced by conductors carrying electric currents. The analogy between velocity fields of vortex filaments and magnetic fields produced by electric currents is so great that many theorems and examples in electrodynamics can be applied directly to hydrodynamics by simply replacing electric currents by vortex filaments and magnetic fields by velocity distributions.

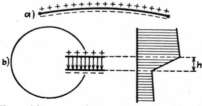

Fig. 132.—(a) The arbitrary surface may be considered as a double layer of sources and sinks. (b) The compensating potential field of the double layer surface.

We shall now consider the problem of obtaining the potential field due to a given vortex. The well-known method of the doublet surface of electrodynamics offers us an opening. We imagine the arbitrary surface (Fig. 132a) to be uniformly covered with sources $(+ + + +)$ on one side and with sinks $(- - - -)$ on the other, and now we assume that instead of the surface we have a double layer of small but finite thickness h (Fig. 132b).

This does away with the difficulty of the discontinuity of potential. By far the largest part of the stream coming from the sources flows directly to the sinks and produces a large difference of potential in the short distance h. The strengths of the sources and sinks are so chosen that the difference of potential is everywhere equal to Γ. Only a proportionately small quantity of the source stream flows out and builds up the rest of the field.

If Q is the strength of a point source, then the value of the velocity $\mathbf{w}$ at a distance a from the source is

$$|\mathbf{w}| = w = \frac{Q}{4\pi a^2},$$

so that the potential of the source is

$$\Phi = \int_0^a w \, da = \text{const.} - \frac{Q}{4\pi a},$$

and that of a sink

$$\Phi = \text{const.} + \frac{Q}{4\pi a}.$$

The potential at a point P due to the element dA of the doublet sheet is

$$d\Phi = -\frac{q \, dA}{4\pi a_+} + \frac{q \, dA}{4\pi a_-} = \frac{q \, dA}{4\pi}\left(\frac{1}{a_-} - \frac{1}{a_+}\right),$$

where q is the source intensity per unit surface, and a_+ and a_- the distances from P to the source and sink elements (Fig. 133). We see, however, that

$$a_+ = a_- + h \cos \alpha,$$

and, if h is small compared to a,

$$\frac{1}{a_+} = \frac{1}{a_-} - \frac{1}{a^2} h \cos \alpha$$

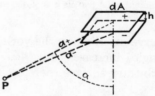

giving

$$\Phi = \frac{q \, dA}{4\pi a^2} h \cos \alpha.$$

FIG. 133.—The potential produced at P due to an area dA of the double layer.

The velocity inside the double layer is equal to the source intensity per unit area q, since but for an infinitesimal portion of the stream all the flow is in the doublet sheet itself. The increase in potential is therefore qh. This however must be equal to Γ if h converges to zero. Therefore $qh = \text{constant} = \Gamma$. Hence

$$d\Phi = \frac{\Gamma}{4\pi} \frac{dA \cos \alpha}{a^2}.$$

The quantity $dA \cos \alpha/a^2$ however is equal to $d\omega$, the solid angle subtended by the surface dA at P. It is equal to the area cut out on a unit sphere with its center at P by the pencil of rays joining the boundary of dA to P.

Finally if we complete the integration for the whole surface **A** which is bounded by the vortex line, we find that the potential at P is

$$\Phi_P = \frac{\Gamma}{4\pi}\omega,$$

where ω is the solid angle subtended by the closed vortex line at the point in question (Fig. 134). If P moves round a closed

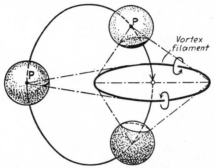

Fig. 134.—Change in value of the solid angle subtended at A by the vortex filament, as P travels in a closed path which cuts the arbitrary surface containing the filament.

curve interlinking the vortex ring, as shown in Fig. 134, the solid angle increases by 4π. This closed circuit piercing the arbitrary surface therefore produces a change of

$$\frac{\Gamma}{4\pi}4\pi = \Gamma$$

in the potential.

The velocity distribution is obtained by taking the gradient of Φ, and, since Γ is constant, this is

$$\mathbf{w} = \operatorname{grad} \Phi = \frac{\Gamma}{4\pi} \operatorname{grad} \omega.$$

In order to calculate grad ω we displace in Fig. 135 the point P to P' by the vector $d\mathbf{r}$ and obtain the change in solid angle

$$d\omega = d\mathbf{r} \circ \operatorname{grad} \omega.$$

Since only the motion of *P relative* to the vortex field is of importance, we can just as well displace the vortex by the same amount in the opposite direction. We now consider the change effected in the solid angle ω when the vortex filament **C** is displaced a distance $d\mathbf{r}$ parallel to itself (Fig. 136). The change is obviously the projection of the cylindrical surface **CC'** on the unit sphere. If $d\mathbf{s}$ is an element of the filament **C**, then an element of area of the cylindrical surface is $d\mathbf{r} \times d\mathbf{s}$ and its projection on a plane at right angles to the radius vector a is

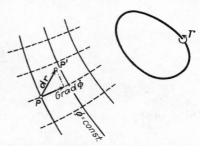

$$d\mathbf{r} \times d\mathbf{s} \circ \frac{\mathbf{a}}{a} = d\mathbf{r} \circ d\mathbf{s} \times \frac{\mathbf{a}}{a};$$

FIG. 135.—Displacement of the point P through a distance $d\mathbf{r}$.

the projection on the unit sphere is

$$\frac{d\mathbf{r} \circ d\mathbf{s} \times \frac{\mathbf{a}}{a}}{a^2}.$$

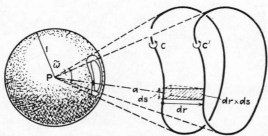

FIG. 136.—Change of solid angle subtended by the closed vortex line **C** at P, by moving the vortex line a distance $d\mathbf{r}$ parallel to itself.

Hence the projection of **CC'** on the unit sphere, *i.e.*, the change of solid angle, is

$$d\omega = \oint^{\mathbf{C}} d\mathbf{r} \circ \frac{d\mathbf{s} \times \mathbf{a}}{a^3}$$

and since $d\mathbf{r}$ is constant (parallel displacement),

$$d\omega = d\mathbf{r} \circ \oint^{\mathbf{C}} \frac{d\mathbf{s} \times \mathbf{a}}{a^3}.$$

But we know that $d\omega = d\mathbf{r}$ grad ω, so it follows that

$$\text{grad } \omega = \oint^C \frac{d\mathbf{s} \times \mathbf{a}}{a^3}.$$

Inserting this expression into the equation

$$\mathbf{w} = \frac{\Gamma}{4\pi} \text{ grad } \omega,$$

we get, finally,

$$\mathbf{w} = \frac{\Gamma}{4\pi} \oint^C \frac{d\mathbf{s} \times \mathbf{a}}{a^3}.$$

The expression $d\mathbf{s} \times \mathbf{a}$ has the value $ds\, a \sin (ds, a)$ and has a direction perpendicular both to $d\mathbf{s}$ and to $\mathbf{a}$. The velocity $\mathbf{w}$ is obtained by adding together the contributions of the individual

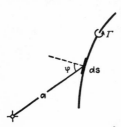

FIG. 137.—A curved vortex line in a plane.

filament elements $d\mathbf{s}$, and the contribution of this element is perpendicular to $d\mathbf{s}$ and $\mathbf{a}$, is proportional to the sine of the angle between $d\mathbf{s}$ and $\mathbf{a}$, and is inversely proportional to the square of the distance a from the point in question. This however is exactly the law of Biot and Savart in electrodynamics from which the magnetic field in the neighborhood of a current-carrying wire can be calculated. If the vortex filament lies in a plane (Fig. 137), the above equation simplifies to

$$w = \frac{\Gamma}{4\pi} \oint^C \frac{ds}{a^2} \sin \varphi.$$

As an example of this latter case we shall calculate the velocity field caused by an infinite rectilinear vortex of circulation Γ (Fig. 138). The previous equation gives

$$w_P = \frac{\Gamma}{4\pi} \int_{-\infty}^{+\infty} \frac{ds \sin \alpha}{a^2}.$$

Considering the angle α as the variable, we have

$$-s = R \cot \alpha, \qquad ds = \frac{R\, d\alpha}{\sin^2 \alpha}, \qquad a = \frac{R}{\sin \alpha}.$$

These give

$$w_P = \frac{\Gamma}{4\pi R} \int_0^\pi \sin \alpha\, d\alpha = -\frac{\Gamma}{4\pi R} \cos \alpha \Big|_0^\pi = \frac{\Gamma}{2\pi R}.$$

This same result can be obtained in a simpler way by drawing a plane through P at right angles to the vortex filament and tak-

ing the line integral in this plane round a circle. We then have

$$\Gamma = \oint^C \mathbf{w} \circ d\mathbf{r},$$

or, since $\mathbf{w} \| d\mathbf{r}$, we can write

$$\Gamma = \oint^C w\,dr = w2\pi R,$$

giving

$$w = \frac{\Gamma}{2\pi R}.$$

The previous calculation, however, can still be applied when the vortex filament is composed of several straight stretches. It is then only necessary to put the correct limits in the corresponding integrals.

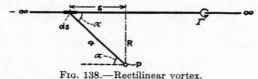

Fig. 138.—Rectilinear vortex.

89. Simplified Construction of a Vortex Line by Assuming a Core of Constant Rotation.—For many calculations it is quite sufficient to replace the vortex by the simplified structure to be described below.

In isolated vortex filaments the rotation is usually greatest at the center and gradually decreases to zero on the boundary. In calculations, however, it is usual to assume a vortex core of constant rotation (so that a rectilinear vortex may also be considered as a rigid rotating body) and a potential motion outside the core (Fig. 139). The velocity on the circumference $\mathbf{w}_1$ of this imaginary rigid body is connected with the angular velocity q by the relation $w_1 = qr_1$, where r_1 is the radius of the core.

Outside the core a potential flow exists in which the velocity at a distance r from the center is $w = \Gamma/2\pi r$ (see Fig. 140). This circulation Γ is equal to the strength of the vortex:

$$\Gamma = |\text{rot } \mathbf{w}| \cdot \pi r_1^2 = 2q \cdot \pi r_1^2.$$

From another point of view we get

$$\Gamma = 2\pi r_1 \cdot |\mathbf{w}_1| = 2\pi r_1^2 q$$

which agrees with the above.

With this value of Γ the velocity in the region of potential flow is

$$w = \frac{\Gamma}{2\pi r} = q\frac{r_1{}^2}{r}.$$

In case $r = r_1$, the velocity of the potential flow is equal to that of the core. The components u and v in the x- and y-directions, respectively, are

$$u = -|\mathbf{w}|\frac{y}{r}, \qquad v = |\mathbf{w}|\frac{x}{r}.$$

In the core these values are

$$u = -qy, \qquad v = qx;$$

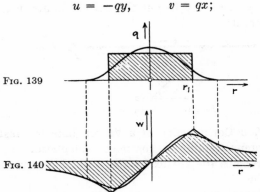

Fig. 139.

Fig. 140.

Fig. 139.—Simplified construction of a vortex filament by assuming a vortex core of constant rotation.

Fig. 140.—Distribution of velocity due to a vortex filament and that due to a vortex core surrounded by potential flow.

and in the potential flow,

$$u = -q\frac{r_1{}^2}{r^2}y, \qquad v = q\frac{r_1{}^2}{r^2}x.$$

The velocity distribution expressed by these equations is shown in Fig. 140 by the shaded area.

90. The Motion and Mutual Influence of Single Vortices.—In discussing the motion of individual vortices we start from the Helmholtz theorem stating that each vortex in its motion is always composed of the same fluid elements. It follows that if an extraneous velocity field is superposed on the field of the vortex itself, the motion of vortex will be governed by that of the extraneous field. Such a state of affairs exists when for example one vortex is influenced by the field of another vortex.

As an example we shall consider two infinite parallel rectilinear vortices of strengths Γ_1 and Γ_2, rotating in the same sense. The potential fields and therefore the velocity distributions of these two systems are superposable. In Figure 141, let $a = a_1 + a_2$ be the distance between the two vortices; then the velocity at point 2 is that due to vortex Γ_1 and is equal to

$$w_2 = \frac{\Gamma_1}{2\pi a},$$

and that at point 1 is due to the vortex Γ_2, its value being

$$w_1 = \frac{\Gamma_2}{2\pi a}.$$

Both these velocities are perpendicular to a; and when Γ_1 and Γ_2 rotate in the same sense, the velocities are opposed to each other.

If each vortex is given a mass equal to its strength, we can define a "center of gravity" S of the vortex system which lies on the line joining the vortices and is given by the relation $\Gamma_1 a_1 + \Gamma_2 a_2 = 0$. The velocity of the center of gravity is now

$$w_S = \frac{\Gamma_1 w_1 + \Gamma_2 w_2}{\Gamma_1 + \Gamma_2}.$$

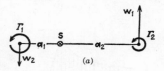

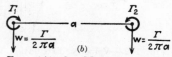

Fig. 141*a–b.*—Mutual influence of two infinite rectilinear parallel vortices. (*a*) Vortices have different strengths of same sign. (*b*) Vortices have same strengths but of opposite sign.

On evaluation this is seen to be zero, so that the center of gravity of the system is always at rest. This only means that the vortices always have the same space point as the center of their circular paths; nothing at all is said about the fluid velocity at this point and it is by no means necessary that it should be zero.

Analogous superpositions are true for any number of parallel vortices. We can say that every vortex filament moves in the path prescribed by the other vortices. This statement, as well as the center of gravity theorem given above, is due to Helmholtz.

As a further special case let us consider the motion of two vortices of equal and opposite circulation, $\Gamma_1 = -\Gamma_2$. The center of gravity of this system is at infinity so that the "vortex pair" has a straight-line motion (Fig. 141). The vortex Γ_1

produces a velocity $\Gamma_1/2\pi a$ on vortex Γ_2 at right angles to the line joining them. In addition the vortex Γ_2 induces a velocity, $-\Gamma_2/2\pi a$, on vortex Γ_1, also at right angles to a. Hence, since $\Gamma_1 = -\Gamma_2$, the vortex pair moves with constant velocity $\Gamma/2\pi a$ at right angles to the joining line. The velocity at the mid-point of these two vortices is

$$\frac{\Gamma}{2\pi\frac{a}{2}} + \frac{\Gamma}{2\pi\frac{a}{2}} = \frac{2\Gamma}{\pi a}$$

which is four times that of the individual vortices.

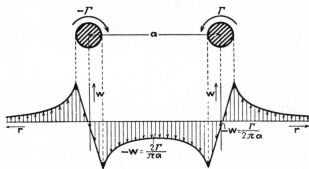

Fig. 142.—Velocity distribution in the neighborhood of a rectilinear vortex pair.

The fluid particles on the joining line have velocities relative to the vortices given by

$$w = \frac{\Gamma}{2\pi r_1} - \frac{\Gamma}{2\pi r_2}$$

where r_1 and r_2 are measured from the mid-points of the two vortices. This expression for the velocity field is true only in the region outside the cores, as shown in Fig. 142.

In order to derive the streamlines and potential lines it is perhaps best to return to the representation by means of a complex argument. From Art. 78 the stream function for a vortex is

$$F = i \log z = \Phi + i\Psi.$$

The corresponding function for two equal and opposite vortices at right angles to the z-plane and cutting it at z_1 and z_2, respectively, is

$$F = \frac{\Gamma}{2\pi}i[\ln (z - z_1) - \ln (z - z_2)]$$

or

$$\Phi + i\Psi = \frac{\Gamma}{2\pi}(\varphi_2 - \varphi_1) + i\frac{\Gamma}{2\pi}\ln\frac{r_1}{r_2}.$$

The curves of constant potential therefore are lines for which $\varphi_2 - \varphi_1$ is constant. These curves are a family of circles through z_1 and z_2 as shown in Fig. 143.

The curves of constant Ψ (the streamlines) are the family of curves orthogonal to this system, and they are given by the family of circles r_1/r_2 = constant. However, it is important to remember that this flow is not steady, for the vortices themselves are in motion. In order to obtain the steady state we must reduce to zero the velocity of the vortices, which we found to be $\Gamma/2\pi a$. The corresponding streamline pattern is shown in Fig. 144. In the infinite part of the plane the velocity is $\Gamma/2\pi a$, and at

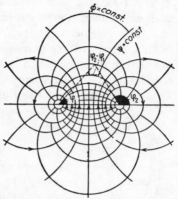

Fig. 143.—Stream lines and lines of constant potential for two rectilinear vortices of equal and opposite circulation. (This is non-steady motion since the vortex pair possesses a velocity).

the mid-point of the line joining the vortices the velocity is $3\Gamma/2\pi a$ in the opposite direction. As seen in Fig. 144, two types of

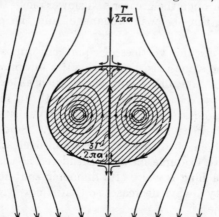

Fig. 144.—The steady motion corresponding to the case of two rectilinear vortices of equal and opposite circulation as shown in Fig. 143.

streamlines are formed, a closed system and an open system. Physically this means that the liquid in the interior of the closed

streamline accompanies the vortices in their flow and the approaching stream flows around as if there were a rigid body in the fluid. The shaded portion in the diagram represents the "body."

A vortex ring has some similarity to a vortex pair in that each element of the ring is influenced by the remaining elements and thus a velocity is given to the whole ring. The integration in this case is rather more involved. We find that the velocity of the ring is greater than that of the vortex pair, and the smaller the diameter of the vortex core the greater is the velocity. If D is the diameter of the ring and d that of the core (Fig. 145), then the velocity of the ring is

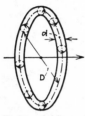

FIG. 145.—A vortex ring.

$$\frac{\Gamma}{\pi D}\left(\ln \frac{8D}{d} - \frac{1}{4}\right).$$

If there are several vortex rings in the field it is clear that they will influence each other's motion. For example, with two circular vortices of the same sense of circulation following each other along a common straight axis, the mutual effect makes itself noticeable as follows. The vortex in front increases its radius, thus causing its velocity to diminish. The one in the rear contracts and therefore proceeds with increased velocity, till finally the smaller and speedier vortex passes right through the greater and slower one. The rôles of the vortices have thus been interchanged, and the process repeats itself.

Two equal circular vortices of equal and opposite circulation, moving on the same axis, have velocities in opposite directions. They influence each other however to such an extent that they continually draw closer to each other with an increase in their radii. This produces a large decrease of velocity, so that in a finite time they never touch each other. Since the plane of symmetry, which both these vortices approach, acts like a rigid wall, the above example may be considered as that of a circular vortex approaching a wall. Its diameter continually increases, but its velocity always decreases so that it never reaches the wall in a finite time.

The type of fluid "body" carried along by the circular vortex has also been investigated. By superposing a velocity equal and opposite to that of the vortex, the flow is reduced to the

steady state as in the case of the vortex pair, and we find that different diameters of the core produce different types of fluid body as shown in Figs. 146 to 148. In Fig. 146 the fluid flows past a rather thick circular vortex just as if there were a solid body (shaded portion) in the liquid. Figure 147 shows how the shape of the obstacle changes as the vortex becomes thinner, and in Fig. 148, where the vortex is very thin, we simply have a ring traveling along through the fluid.

If such a vortex ring is produced in a dyed fluid and allowed to enter an undyed medium, the forms shown in Figs. 146 to 148 become visible (smoke rings, etc.). The technique whereby

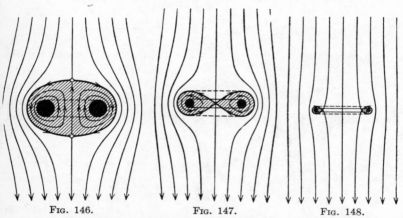

Fig. 146. Fig. 147. Fig. 148.

Figs. 146–148.—Steady streamlines due to vortex rings of cores of various thicknesses.

such vortex filaments or vortex rings can be formed is described in Arts. 71 and 93.

91. Pressure Distribution in the Neighborhood of a Rectilinear Vortex.—The simplifying assumption of the existence of a vortex core enables us to calculate the pressure distribution quite readily. Since we are dealing with a steady motion, Euler's equation gives

$$|\mathbf{w} \circ \nabla \mathbf{w}| = \frac{|\text{grad } p|}{\rho}.$$

In circular motion

$$|\mathbf{w} \circ \nabla \mathbf{w}| = \frac{w^2}{r} \text{ (central acceleration)}.$$

The pressure increases radially outward and therefore

$$\frac{1}{\rho}\frac{dp}{dr} = \frac{w^2}{r}.$$

If the pressure at infinity is p_0, then at any finite point it is

$$p = p_0 - \int_r^\infty \rho\frac{w^2}{r} \cdot dr.$$

With $w = \Gamma/2\pi r$, the integral becomes

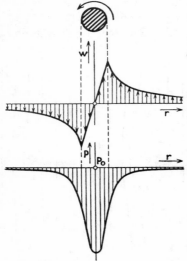

$$\rho\int_r^\infty \frac{w^2}{r} \cdot dr = \frac{\rho\Gamma^2}{4\pi^2}\int_r^\infty \frac{dr}{r^3} = \frac{\rho\Gamma^2}{8\pi^2 r^2}.$$

which agrees with the Bernoulli equation

$$p = p_0 - \frac{\rho w^2}{2}.$$

The pressure on the surface of the core therefore is

$$p_1 = p_0 - \frac{\rho\Gamma^2}{8\pi^2 r_1^2}.$$

Inside the core

$$w = \frac{\Gamma r}{2\pi r_1^2}$$

giving

$$\rho\int_r^{r_1}\frac{w^2}{r}dr = \frac{\rho\Gamma^2}{4\pi^2 r_1^4}\int_r^{r_1} rdr = \frac{\rho\Gamma^2(r_1^2 - r^2)}{8\pi^2 r_1^4}.$$

Fig. 149.—Velocity and pressure distribution in the interior, and in the neighborhood, of a rectilinear vortex.

Hence the pressure in the core is

$$p = p_1 - \rho\int_r^{r_1}\frac{w^2}{r}dr = p_0 - \frac{\rho\Gamma^2(2r_1^2 - r^2)}{8\pi^2 r_1^4}\text{(paraboloid)}.$$

When $r = 0$, *i.e.*, at the center,

$$p_c = p_0 - \frac{\rho\Gamma^2}{4\pi^2 r_1^2}$$

or

$$p_0 - p_c = 2(p_1 - p_0).$$

We therefore see that in the core, where there is rotation, the Bernoulli equation is no longer applicable.

Figure 149 shows the velocity and pressure distributions.

92. The Relation between Vortex Motion and the Surface of Discontinuity or Separation.—We noted in Art. 83 that the

existence of rotation in the interior of the fluid could be explained by causes other than the arrival of particles from the boundary layer. We refer to the so-called surface of discontinuity which has been the subject of much investigation, particularly by Helmholtz and Kirchhoff.

Even though it is still impossible to enter into all the details of the theory of this phenomenon, it is of importance to explain at least how vortex motion is produced from surfaces of discontinuity. The reasons for the formation of surfaces of discontinuity may be many.

Let us consider the state of affairs that exists when two fluid layers of different velocities meet at the sharp trailing edge of an obstacle (K, Fig. 150).

Fig. 150.—The surface of discontinuity formed at the sharp trailing edge of the body K, where layers of fluid of different velocities meet, changes into a fluid layer with rotation under the assumption of a small fluid viscosity.

The surface of discontinuity that is thus formed and which takes its name from the fact that there is a discontinuity in velocity across it, is the surface of separation of the two fluid layers. We can imagine each layer marked with a definite color so that the surface of separation becomes visible.

In considering the velocity diagram behind the sharp edge, we see that there is a discontinuity in the velocity, as shown in Fig. 150. At least this is so for non-viscous liquids, because, if we assume even the smallest viscosity, this discontinuity is evened out, and instead of the sudden jump we have the transition spread over a distance.

This transition zone is a layer of rotation as can be seen by the small fluid cross in Fig. 150. Assuming the x-axis in the direction of the velocity and the y-axis at right angles, we find for a two-dimensional motion:

$$\text{rot } \mathbf{w} = \frac{\partial v}{\partial x} - \frac{\partial u}{\partial y};$$

and, since $\partial u/\partial y \neq 0$, but $\partial v/\partial x = 0$, we have

$$\text{rot } \mathbf{w} \neq 0,$$

or, in other words, there is rotation in the transition layer. If the viscosity converges to zero, the transition layer becomes the surface of separation. Although **w** and also Φ are discontinuous at the surface of separation, physical reasons show that the pressure is continuous.

An example of the type of motion that occurs when fluid originally in contact is separated and meets behind the sharp edge is given by the flow round a finite airfoil (three-dimensional motion). Since in Fig. 151 there is no difference between the pressures at A on both sides of the surface of separation, we find on integrating the Bernoulli equation from O along both branches of the streamline that the value of w is also identical on both sides of the surface. The directions of **w,** however, need not be the same. This means that in the steady

Fig. 151.—The surface of separation of an airfoil. The fluid particles, flowing together from both sides of the airfoil to make the surface of discontinuity, were previously in contact with each other.

state transverse but no longitudinal discontinuities of velocity can occur.

In the steady two-dimensional case a transverse motion *i.e.,* a motion in the third dimension is quite impossible; there is no discontinuity in velocity. On the other hand a jump in potential, or, in other words, a circulation round the airfoil, is quite possible (see Art. 80).

In non-steady motion, $\partial\Phi/\partial t$ and therefore also the numerical value of **w** on the surface may be discontinuous.

93. The Formation of Surfaces of Discontinuity.—In Art. 92 it was seen that the connection between vortex motion and the surfaces of discontinuity of potential motion lies in the fact that any small internal friction changes the discontinuity in velocity into a gradual transition in a layer with rotation.

In the domain in which this continuous change takes place we have a layer of vorticity formed out of vortex filaments, whereas outside the layer there is potential flow (Fig. 152). Assuming that the rotation in this layer is constant we get the velocity distribution shown in Fig. 153.

If now we allow the viscosity to decrease in value toward zero, the quantity h also decreases to zero and the vortex layer thus becomes a vortex surface. A surface of discontinuity may therefore be considered as a surface distribution of vortices, *i.e.*, a vortex surface.

Vortices in slightly viscous fluids are almost always generated by the confluence of fluids and the resulting production of surfaces

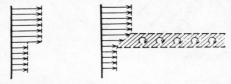

Fɪɢ. 152.—The surface of discontinuity becomes a fluid layer with rotation on changing from the ideal non-viscous liquid to the liquid with internal friction.

of discontinuity. The preliminary cause in the formation of vortices usually lies in the frictional layer directly on the body.

Let us consider the case of two-dimensional flow round a sharp edge as an example of the formation of a surface of discontinuity in non-steady motion. At the commencement of flow the streamlines are almost of the potential type (Fig. 154). As more and more liquid of slow velocity collects in the boundary layer, the flow assumes the shape shown in Fig. 155. Further developments are shown in Figs. 156 and 157 where the confluence at the edge is recognizable. A surface of discontinuity projects

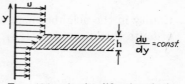

Fɪɢ. 153.—A simplification is introduced into Fig. 152 by assuming that the change of velocity through the layer is linear. A fluid layer of constant rotation is thus obtained.

into the liquid. On either side of this surface the flow is of the potential type, but in the surface itself there is a discontinuity in velocity.

The derivation of this potential introduces great difficulties and up to the present has been possible in only very few simple cases. These difficulties arise from the fact that two conditions must always be satisfied on the surface of separation. First of all this surface must be a fluid surface since it can be considered as a vortex surface, and secondly the pressure must be continuous.[1]

[1] See paper by L. Prandtl: On the Generation of Vorticity in an Ideal Fluid with Applications to Airfoil Theory and other Problems (German); *Papers on Hydro- and Aerodynamics* at Innsbruck, Berlin, 1923.

It is of interest to know the amount of vorticity generated per second. To find this, we decompose the surface of discon-

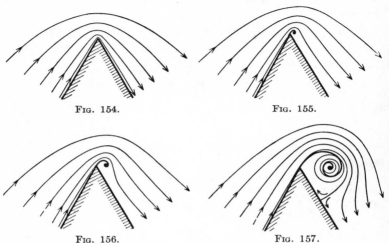

Fig. 154.

Fig. 155.

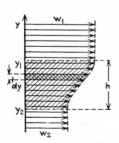

Fig. 156.

Fig. 157.

Figs. 154–157.—Different stages in the formation of a vortex from a non-viscous fluid through flow round a sharp edge.

tinuity in the neighborhood of the sharp edge into a vortex layer of finite thickness (Fig. 158). In this layer

$$|\text{rot } \mathbf{w}| = \frac{dw}{dy}.$$

In order to find the rate of generation of vorticity we have to determine the sectional area, filled with vorticity of strength $|\text{rot } \mathbf{w}|$ that flows by per second. An element of the layer of depth dy has velocity w, so that the cross-sectional area per second of the vortex is

$$\int_{2}^{y_1} w \, dy,$$

and the rate of flow of vorticity is

$$\int_{y_2}^{y_1} w \, dy \cdot \frac{dw}{dy} = \frac{w^2}{2}\bigg|_{y_2}^{y_1} = \frac{w_1{}^2 - w_2{}^2}{2}.$$

Fig. 158.—Distribution of velocity in a layer of separation.

This expression shows that the vorticity generated per second is not dependent on the thickness of the layer h so that we can proceed to the limit $h = 0$ without altering the above result.

As soon as the velocities in the non-steady example become equal on both sides of the surface of separation, *i.e.*, as soon as

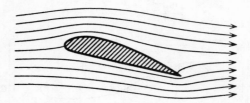

Fig. 159.—Streamline form in the first instant of flow round the airfoil. (Potential flow).

we can put $w_1 = w_2$ in the expression for the quantity of vorticity generated, no more vortices are produced.

This occurs for example in the case of an airfoil. If the airfoil proceeds with constant velocity through the air, or if a stream of constant velocity flows against it, the streamline pattern in

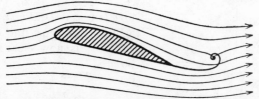

Fig. 160.—Formation of the vortex—the streamline pattern some time after that of Fig. 159.

the first instant of motion is almost of the potential type as shown in Fig. 159. After a short time the form of the streamlines changes to that shown in Fig. 160. The strength of vorticity increases till $w_1 = w_2$ (Fig. 161).

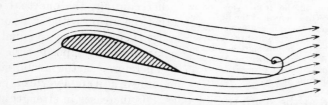

Fig. 161.—The vortex grows till the values of the velocities above and below the surface of separation are equal.

In this position the circulation round the airfoil is constant and is equal and opposite to that of the departed vortex. If we

take a fluid line $ABCD$, as in Fig. 162, sufficiently big to encompass the airfoil and the vortex, then obviously

$$\Gamma = \oint^{ABCDA} \mathbf{w} \circ d\mathbf{s} = 0,$$

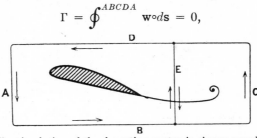

Fig. 162.—The circulation of the departing vortex is always equal and opposite to that round the airfoil.

for $ABCDA$ was a closed fluid curve when the liquid was at rest, and according to Thomson's theorem the line integral of the velocity along a closed fluid curve is constant for all time.

The foregoing assertion can then be proved by splitting the whole region into two parts, $ABEDA$ and $DEBCD$, by inserting the transverse line BED (Fig. 162). It is now clear that

$$\oint^{ABCDA} \mathbf{w} \circ d\mathbf{s} =$$

$$\oint^{ABEDA} \mathbf{w} \circ d\mathbf{s} +$$

$$\oint^{EBCDE} \mathbf{w} \circ d\mathbf{s} = 0.$$

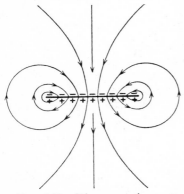

Fig. 163.—Diagrammatic representation of the front view of an airfoil. On account of the difference of pressure there is a flow round the sides. (Smaller pressure above—shown by the minus sign, and bigger pressure below—shown by the positive sign.) The figure is drawn to scale for fluid at rest at infinity.

The second contour surrounds the vortex and is therefore certain to possess a circulation. The other contour, the one that contains the airfoil without a vortex, must therefore have an equal circulation but of opposite sign.

Therefore, as the airfoil starts moving in still air, a vortex is being formed at the sharp trailing edge. It increases in strength until w_1 becomes equal to w_2. From this instant no more circulation is developed, the vortex flows downstream, and the motion becomes steady. This is the explanation of the formation of circulation round a body.

The total circulation round the airfoil is equal to the algebraic sum of the circulations of the vortices that have left and passed downstream. As the velocity of the airfoil varies, vortices of both directions of rotation may be given off. We saw in Art. 80 that the circulation is intimately connected with the lift, in fact a lift is possible in potential flow only when a circulation is introduced.

As an example of the formation of surfaces of discontinuity in cases of steady flow, let us consider the flow round an airfoil of finite length (three-dimensional problem). An airfoil which

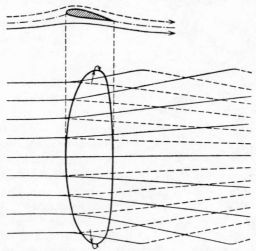

FIG. 164.—Flow above (continuous lines) and below (broken lines) the surface of discontinuity stretching behind the airfoil.

produces a lift must have low pressure above it and high pressure on its bottom surface. As a result of this difference in pressure there is a leakage flow round the edges from the lower face to the upper, as shown in Fig. 163 which is a front view of the airfoil. The region of low pressure is marked $----$, and the high pressure area is shown by $++++$. The leakage flow combined with the main flow of air from front to rear gives a streamline picture somewhat as it is shown in Fig. 164. It is seen that a surface of separation with a transverse discontinuity of velocity leaves the trailing edge.

94. Instability of the Surface of Discontinuity.—Actual surfaces of separation have shapes which differ from those suggested

by the theory, and the reason for this is that these surfaces are unstable. Any disturbance however small (which naturally always occurs) grows with time and soon alters the whole shape of the surface.

The calculations are similar to those connected with water waves. With the complex variable $z = x + iy$, we can write for the stream functions for the regions above and below the surface of separation:

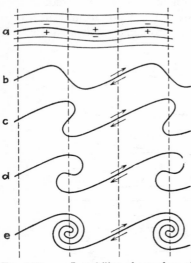

$$F_1(z) = A_1 z + B_1 e^{i\alpha z} e^{\beta t},$$
$$F_2(z) = A_2 z + B_2 e^{-i\alpha z} e^{\beta t}$$

where the first terms are the main flow and the second terms express the superposed small disturbance.

We shall not go into the details of the calculation here, but in carrying it out we first express the condition that the pressure is continuous over the separating surface and that the two fluids do not overlap or leave any empty spaces between them.

When the densities of the two fluids above and below are the

FIG. 165*a–e.*—Instability of a surface of separation.

same and when the only discontinuity is in velocity, we find that β is always complex and that the real part of one of its values is positive (the real part of the second value is negative). The first mentioned value of β always produces instability. These investigations, which are valid only for small motions, were put forward by Lord Rayleigh. He also discussed the case of layers of separation as shown in Fig. 153. Qualitatively the motion is somewhat as follows: Fig. 165*a* shows a small wave-like disturbance on the surface of separation. In the lower region, when the flow is steady, the velocity is increased in the troughs and decreased in the crests. These regions are respectively places of lowered pressure $(-)$ and increased pressure $(+)$. This condition is exactly reversed in the upper layer, and the pressure differences are such as to increase the amplitude of the wave disturbance.

The contour of the wave becomes more and more distorted and is soon unsymmetrical (Fig. 165*b,c*). The waves finally overlie each other and coil up like vortices (Fig. 165*d,e*). The vortices themselves are usually in a position of unstable equilibrium and under their mutual influences they soon completely disarrange the streamline pattern.

CHAPTER XIII

THE INFLUENCE OF COMPRESSIBILITY

95. General Remarks about the Justification for Treating Gases as Incompressible Fluids.—With the exception of Chaps. II, III, and IV we have so far discussed in this book the laws of motion of "fluids," without differentiating between liquids and gases. We neglected the fact that gases, as opposed to liquids, are compressible and can have a density which varies from point to point.

Since density plays an important part in the equations of motion, a change of density will produce a change in the type of flow. In this chapter we shall show that, in the majority of cases, liquids and gases obey the same laws of motion and that gases can generally be treated as liquids, in spite of the fact that their compressibility is essentially different.

Neglecting the effects of absorption or emission of heat, changes in the density of a gas can be produced only by changes in pressure. Since the connection between the density ρ (and therefore the specific volume $v = 1/g\rho$) and the pressure in the adiabatic state is expressed by the equation[1]

$$pv^k = \text{const.},$$

it follows that a 1 per cent change of density is produced by a 1.405 per cent change of pressure. When the air is under atmospheric pressure (14.4 lb/in.²), this is equivalent to a pressure change of 0.19 lb/in.² We therefore perceive that the density changes induced by forces of this magnitude are very small and it will be explained later that they have hardly any effect on the type of flow of the gas.

Changes of density can be produced in the following three ways:

1. Artificially created pressure changes in vessels and pipes.
2. Great differences of height.
3. Very great velocities.

---

[1] k is the ratio of specific heat at constant pressure (c_p) to that at constant volume (c_v): [$k = c_p/c_v$] and its value for air at atmospheric pressure is 1.405.

This chapter will show the effect of compressibility on the kinematics and dynamics of gases and will determine the limiting heights and velocities above which compressibility cannot be neglected without introducing appreciable errors.

96. Compressibility in Bernoulli's Equation.—In order to determine the effect of compressibility on gases we have to investigate the integral

$$P = \int_{p_1}^{p} \frac{dp}{\rho}$$

which arises from Bernoulli's equation for steady motion

$$\frac{w^2}{2} + gh - P = \text{const.}$$

With $\rho = \gamma/g = 1/gv$, where $v = 1/\gamma$ is the volume of unit weight, this integral becomes:

$$P = g \int_{p_1}^{p} v \, dp.$$

Since the usual types of flow deal with adiabatic changes of state, the relation between v and p is given by the equation

$$pv^k = \text{const.}$$

Let the specific volume corresponding to p_1 be v_1, and we get

$$pv^k = p_1 v_1^k$$

or

$$v = \frac{p_1^{(1-k)/k}}{p^{1/k}} \cdot p_1 v_1,$$

and, by introducing the height of the uniform atmosphere $h_0 = p_1 v_1$ (see page 30),

$$v = h_0 p_1^{(1-k)/k} \cdot p^{-1/k}.$$

Inserting this expression into the equation for P, we have

$$P = gh_0 p_1^{(1-k)/k} \int_{p_1}^{p} \frac{dp}{p^{1/k}} = gh_0 p_1^{(1-k)/k} \cdot \frac{k}{k-1} p^{(k-1)/k} \Big|_{p_1}^{p} =$$
$$gh_0 \frac{k}{k-1} \left\{ \left(\frac{p}{p_1} \right)^{(k-1)/k} - 1 \right\}.$$

We now introduce this value into Bernoulli's equation and on dividing by g we obtain

$$\frac{w^2}{2g} + h + h_0 \frac{k}{k-1} \left\{ \left(\frac{p}{p_1} \right)^{(k-1)/k} - 1 \right\} = \text{const.}$$

As a special case, consider $w = 0$, corresponding to the static equilibrium. Let the height h_1 correspond to the pressure p_1 and we have

$$h - h_1 = h_0 \frac{k}{k-1} \left\{ 1 - \left(\frac{p}{p_1} \right)^{(k-1)/k} \right\}. \tag{1}$$

This is the equation obtained in Art. 13 (Eq. 9) for the adiabatic distribution.

If the height remains constant in the problem under investigation, or if we are dealing only with small differences of height, $(h - h_1)$ can be neglected or absorbed in the constant of Bernoulli's equation, leaving us with a relation between the velocity w and the pressure p:

$$\frac{w^2}{2g} + h_0 \frac{k}{k-1} \left\{ \left(\frac{p}{p_1} \right)^{(k-1)/k} - 1 \right\} = \text{const.} \tag{2}$$

The order of the height below which this last equation is of practical value is given by a statement made in Art. 11 where it was said that a depth of 260 ft in a uniform atmosphere only corresponds to a pressure variation of 1 per cent. The usual problems that come before us (with the exception of those of meteorology) deal with heights well below this value, so that the above equation is valid in most practical cases where no undue accuracy is required.

We shall consider two examples. The constant in Eq. (2) is arbitrary so that the velocity corresponding to p_1 is still undefined. Let us assume that p_1 is the pressure at rest ($w = 0$), so that the constant is zero. Again inserting $h_0 = p_1 v_1$ in (2), we obtain the equation for the velocity of efflux of gas from a container:

$$\frac{w^2}{2g} = p_1 v_1 \frac{k}{k-1} \left\{ 1 - \left(\frac{p}{p_1} \right)^{(k-1)/k} \right\},$$

and therefore

$$w = \sqrt{2g p_1 v_1 \frac{k}{k-1} \left\{ 1 - \left(\frac{p}{p_1} \right)^{(k-1)/k} \right\}} = \sqrt{2g(h - h_1)}, \tag{3}$$

where $(h - h_1)$ is the height difference from Eq. (1) corresponding to the pressure change $(p_1 - p)$ by which the velocity w was generated. The greatest velocity is obtained when $p = 0$, which gives

$$w_{\text{max.}} = \sqrt{2g \frac{k}{k-1} p_1 v_1}.$$

In order to calculate this value numerically for $p_1 = $ atmospheric pressure, we observe that

$$p_1 v_1 \frac{k}{k-1} = h_0 \frac{k}{k-1} = h^\star,$$

in which $h^\star$ is the height of the adiabatic atmosphere,

$$h^\star = \frac{1.405}{0.405} \times 5 \text{ miles} = 92,000 \text{ ft.}$$

Thus the velocity of efflux from atmospheric pressure into a vacuum is

$$w = \sqrt{2 \times 32.2 \times 92,000} = 2,400 \text{ ft/sec.}$$

97. The Effect of Compressibility on the Formula for Stagnation Pressure.—Let us now consider the case of a body moving with a definite velocity through still air. Since only the motion of the body relative to the air is of importance we obtain the same relations when the air is considered flowing past the body at rest.

At a sufficient distance from the body, where the flow is not influenced by it, let the velocity be w_0 and the pressure p_0. At the stagnation point, let w_1 be zero and the pressure be p_1.

Inserting p_0 instead of p_1 in Eq. (2), we find that the constant is $w_0^2/2g$; therefore

$$\frac{w_0^2 - w^2}{2g} = h_0 \frac{k}{k-1} \left[\left(\frac{p}{p_0} \right)^{(k-1)/k} - 1 \right].$$

Putting $w = 0$ and $p = p_1$, we obtain the following equation for the pressure at the stagnation point:

$$\frac{w_0^2}{2g} = h_0 \frac{k}{k-1} \left[\left(\frac{p_1}{p_0} \right)^{(k-1)/k} - 1 \right].$$

Let us now make use of the relation

$$h_0 = p_0 v_0 = \frac{p_0}{\rho_0 g}$$

and solve the equation for p_1. This gives

$$p_1 = p_0 \left(1 + \frac{\rho \dfrac{w_0^2}{2}}{p_0} \cdot \frac{k-1}{k} \right)^{k/k-1}$$

and, denoting the stagnation pressure $p_1 - p_0$ by q, we get

$$q = p_0 \left\{ \left[1 + \frac{\rho_0 w_0^2}{2 p_0} \cdot \frac{k-1}{k} \right]^{k/k-1} - 1 \right\}.$$

In order to see how this value compares with the simple formula for the stagnation pressure of an incompressible fluid we develop the right-hand side of the equation and get

$$q = p_0\left\{1 + \frac{\rho_0 w_0{}^2}{2p_0} + \frac{1}{2k}\left(\frac{\rho_0 w_0{}^2}{2p_0}\right)^2 + \cdots - 1\right\},$$

$$q = \frac{\rho_0 w_0{}^2}{2}\left(1 + \frac{\rho_0 w_0{}^2}{4kp_0} + \cdots\right).$$

If we make use of the fact that the velocity of sound in an ideal gas is

$$c = \sqrt{\frac{kp_0}{\rho_0}},$$

we find

$$q = \frac{\rho_0 w_0{}^2}{2}\left\{1 + \frac{w_0{}^2}{4c^2} + \cdots\right\}. \tag{4}$$

This is the general formula for the stagnation pressure in which

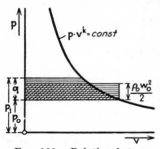

the first term alone gives the corresponding pressure for an incompressible fluid.

It is of interest to investigate at which velocities w_0 the error introduced in Eq. (4) by neglecting the second term in the bracket is less than 1 per cent. These velocities are expressed by

$$\frac{w_0{}^2}{4c^2} \leqq \frac{1}{100}.$$

Fig. 166.—Relation between the stagnation pressure q of a compressible fluid and the stagnation pressure $\rho_0 w_0{}^2/2$ calculated by neglecting compressibility.

Obviously the result is $w_0 \leqq c/5$. If we are dealing with air at 15°C, then $c = 1,120$ ft/sec, so that it is only for speeds above $1,120/5 = 225$ ft/sec that the effect of the compressibility introduces a correction of 1 per cent in the stagnation pressure.

If in the v,p-diagram of Fig. 166 we draw horizontal lines through the points $p = p_1$ (stagnation pressure) and $p = p_0$ (pressure far away), then the area enclosed by these two horizontal lines, by the ordinate axis and by the curve $pv^k =$ constant is equal to the integral $\int_{p_0}^{p_1} v\,dp$. The height $p_1 - p_0 = q$ of this area (which is shaded by horizontal lines) is the stagnation

pressure of the compressible fluid. The corresponding stagnation pressure of an incompressible fluid $\left(\dfrac{\rho_0 w_0{}^2}{2}\right)$ is the height of the rectangular section raised on the line $p = p_0$ between the ordinate axis and the adiabatic curve, such that it has the same area as $\int_{p_0}^{p_1} v\,dp$. In Fig. 166 this new area is shaded with 45-deg lines.

98. Compressibility in the Equation of Continuity.—After having seen how compressibility affects Bernoulli's equation, we shall investigate its influence on the equation of continuity. We start from the theorem expressing the conservation of matter, namely that the mass of fluid flowing through a stream tube of cross section A per second must be constant, *i.e.*,

$$M = A\rho w = \text{const.}$$

For simplicity let us introduce a new quantity a, which is the cross-sectional area of the surface through which unit mass flows in unit time, *i.e.*, $a = A/M$. Our equation of continuity therefore is

$$a w \rho = 1$$

or

$$\frac{1}{a} = w\rho.$$

If the pressure in the fluid at rest ($w = 0$) is p_1 and the corresponding density $\rho = \rho_1$, we obtain the equation

$$\rho_1 w = 0.$$

On the other hand when $p = 0$ (*i.e.*, expansion to complete vacuum), the density ρ is zero. The velocity attained with such an expansion is finite, Eq. (3), so that also

$$\rho w_{\text{max.}} = 0.$$

Between these two extremes $w = 0$ and $w = w_{\text{max.}}$ the quantity $\rho w = 1/a$ is finite and different from zero. Hence the function $1/a$, considered as a function of p, has at least one maximum value between the values of p corresponding to the above two extreme cases, the limits being $p = 0$ and $p = p_1$. Further investigations show that this maximum is the only one in the range (Fig. 167).

If a stream of gas leaves a region of rest ($w = 0$) and enters a domain of smaller pressure or into a vacuum ($p = 0$), the

velocity increases according to Eq. (3). During this process $\rho w = 1/a$ at first increases, *i.e.*, the cross section of the stream tube (a) decreases as the velocity becomes larger. (This of course is also true for incompressible fluids.) The reason for this is that the initial fairly large velocities increase proportionately faster than the decrease of density produced by pressure

$$\frac{dw}{w} > -\frac{d\rho}{\rho}.$$

With a further decrease of pressure and a corresponding increase of velocity the above relation changes sign.   Although a decrease of pressure produces an increase of velocity (which approaches

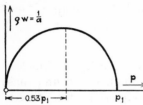

FIG. 167.—Relation between pressure and the reciprocal surface a (through which unit mass flows in unit time).

its asymptotic value for $p = 0$ more and more slowly), the density decreases without limit so that the product ρw finally decreases and attains the value zero.   This means that the area of the stream tube, after having diminished to a minimum for some velocity, now increases with the velocity.

Let us investigate the case where the area of cross section attains its minimum value, *i.e.*, for which ρw is a maximum. At this point

$$\frac{d(\rho w)}{dp} = \rho\frac{dw}{dp} + w\frac{d\rho}{dp} = 0.$$

The term dw/dp will be considered first. By neglecting gravity and assuming a one-dimensional flow (ds is the element of length), Euler's equation is

$$w\frac{dw}{ds} = -\frac{1}{\rho}\frac{dp}{ds}$$

or

$$\frac{dw}{dp} = -\frac{1}{\rho w}.$$

In order to obtain the second term we must use a relation derived in Art. 62:

$$\frac{dp}{d\rho} = c^2$$

or

$$w\frac{d\rho}{dp} = \frac{w}{c^2}.$$

Inserting both these values in the equation

$$\frac{d(\rho w)}{dp} = 0,$$

we get

$$\frac{w^2}{c^2} = 1.$$

We thus see that ρw is a maximum, or the cross section of the stream tube is a minimum, when the velocity w is equal to the velocity of sound, corresponding to the state of the medium at the minimum cross section.

The cross section of the stream tube may decrease with growing velocity as in the case of incompressible fluids or in the case of gases with velocity less than that of sound. It may however increase as when dealing with gases of which the velocity is greater than that of sound. The way in which the area is changing is fundamental and indicates whether the type of flow is that associated with motion above or below the velocity of sound. These types of flow are quite different.

99. The Effect of Compressibility on the Streamlines When the Velocity Is Less than That of Sound.—We begin with Eq. (3), Art. 96, which gives the relation between pressure and velocity. In it we put $v_1 = 1/g\rho_1$ and solve for p, obtaining

$$p = p_1\left(1 - \frac{k-1}{k}\frac{w^2\rho_1}{2p_1}\right)^{k/(k-1)},$$

where p_1 is the pressure in the container corresponding to $w = 0$. We also have

$$pv^k = p_1v_1{}^k$$

or

$$\rho = \rho_1\left(\frac{p}{p_1}\right)^{1/k},$$

giving

$$\rho = \rho_1\left(1 - \frac{k-1}{k}\frac{\rho_1 w^2}{2p_1}\right)^{1/(k-1)}.$$

Expanding the right-hand side, we get

$$\rho = \rho_1\left(1 - \frac{1}{2k}\cdot\frac{\rho_1 w^2}{p_1} + \cdots\right)$$

or, on referring to the relation

$$\frac{kp_1}{\rho_1} = c_1{}^2,$$

where c_1 is the velocity of sound of the gas in the container, we have

$$\rho = \rho_1\left(1 - \frac{1}{2}\frac{w^2}{c_1^2} + \cdots\right).\tag{5}$$

Finally we shall calculate at what velocity the density decreases by 1 per cent. Obviously this occurs when the second term in the series (5) is equal to 0.01 or when

$$w = \sqrt{0.02c_1^2} = \frac{c_1}{10}\sqrt{2}.$$

For air at atmospheric pressure and at a temperature of 15°C the velocity of sound is $c_1 = 1{,}120$ ft/sec, so that

$$w = \frac{1{,}120}{10}\sqrt{2} = 160 \text{ ft/sec.}$$

We thus see that the error in the equation of continuity, or in the form of the streamlines ($\rho w = 1/a$), is greater than the error in the stagnation-pressure formula. In the stagnation-pressure formula we had an error of 1 per cent when the velocity was 225 ft/sec, but in the equation of continuity an error of 1 per cent creeps in when the velocity is only 160 ft/sec.

Equations (4) and (5) show that the error introduced by assuming the gas to be incompressible grows with the square of the velocity, and that the error in the equation of continuity is 4 per cent at speeds of about 300 ft/sec. In many applications a discrepancy of this order is quite tolerable.

CHAPTER XIV

THEOREMS OF ENERGY AND MOMENTUM

100. The Momentum Theorem for Steady Motion.—The undoubted value of theorems on energy and momentum lies in the fact that their application enables one to obtain results in physical problems from just a knowledge of the boundary conditions. There is no need to be told anything about the interior of the fluid or about the mechanism of the motion. These theorems are usually helpful in cases where equations of motion cannot be written down, or at least cannot be integrated, and they give a knowledge of the general flow without going into details. It should also be noted that considerations of energy and momentum provide a useful check in cases where solutions of the differential equations can be obtained.

We shall see that the theorem of momentum is of practical significance only when the flow is steady or when there is a steady "mean" flow, *i.e.*, when a steady general motion can be recognized in an eddying and irregular flow. The theorem of momentum can be applied to cases where there is a loss of energy due to internal friction, but obviously energy considerations cannot be of any use here, for the internal thermal energy produced by friction remains unknown. On the other hand it is just to the non-steady type of flow that energy considerations can be applied, while in steady-state cases (neglecting friction) the energy equation usually yields results of the form zero equals zero.

The theorem of momentum can be obtained in two ways. We can either start from Newton's laws in general mechanics, which has the advantage of giving a good physical understanding; or we can proceed from Euler's equation, which involves a transformation of volume into surface integrals. Let us consider the first method and write down Newton's law stating that the rate of change of momentum ($\Sigma m\mathbf{w}$) of a bounded mass system of discrete particles is equal to the sum of the external forces acting on the system,

233

$$\frac{d}{dt}\left(\sum m\mathbf{w}\right) = \sum \mathbf{P},$$

where $\sum \mathbf{P}$ is the summation over all external forces, *i.e.*, over all forces that are not due to internal action. (The internal forces, *i.e.*, the forces that the particles exert on each other, cancel according to the principle of action and reaction.)

In passing from a system of discrete particles to a fluid—considered as a continuum—the sum $\sum m\mathbf{w}$ changes into the integral $\int \mathbf{w}\,dm = \mathbf{M}$, and the theorem becomes

$$\frac{d\mathbf{M}}{dt} = \frac{d}{dt}\int \mathbf{w}\,dm = \sum \mathbf{P}; \tag{1}$$

or, in words, the rate of change of momentum in a fluid domain bounded by a *fluid surface* is equal to the resultant of the external

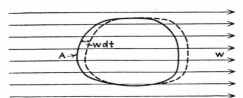

FIG. 168.—The displacement of the fluid mass bounded by the fluid surface **A** in time dt.

forces. It should be noted that the fluid mass under consideration should always be composed of the same elements if Newton's law is to be applied directly. This condition is satisfied if the region is bounded by a fluid surface, *i.e.*, a surface always consisting of the same particles. It is therefore important that during our investigation (differentiation with respect to time) no fluid enters or leaves the boundary that has been defined once and for all.

Let us consider the region bounded by the fluid surface **A** in Fig. 168. A time change in the momentum **M** of this fluid mass can be effected by changes of velocity in the interior of the fluid or by some general motion of the bounding surface (the broken line in Fig. 168 is the position of that surface after a time dt). If the flow is steady, a particle leaving some point in space is replaced by another of equal velocity so that the total change of momentum is that due to the displacement of the fluid surface. This displacement can be interpreted as follows.

Consider a fixed imaginary surface in space, which is coincident with the position of the fluid surface **A** at the time t_1; then, as the fluid surface is displaced, there is a transport of momentum across the fixed surface. In Fig. 169, AA is a portion of the fixed surface which is coincident with that of the moving fluid surface at time $t = t_1$, and BB a part of the fluid surface at time $t_1 + dt$, and, if the normal to the surface is considered positive in the outward direction, the volume of liquid flowing across an area $d\mathbf{A}$ of the surface AA in unit time is $d\mathbf{A} \circ \mathbf{w}dt$ and the momentum sent through the same element is

$$\rho d\mathbf{A} \circ \mathbf{ww}.$$

The total change of momentum in unit time produced by the displacement of the boundary is therefore equal to the resultant of the elemental momenta passing through the fixed surface in this time, i.e.,

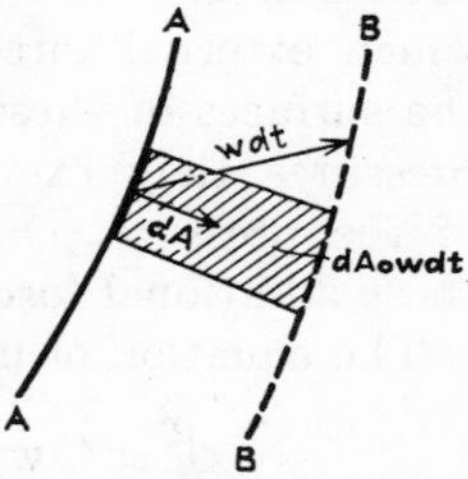

$$\frac{d\mathbf{M}}{dt} = \oiint \rho d\mathbf{A} \circ \mathbf{ww}, \qquad (2)$$

or in components

Fig. 169.—The volume of fluid displaced through the element dA of the non-moving surface in time dt.

$$\frac{dM_x}{dt} = \oiint \rho dA w \cos (\mathbf{n}, \mathbf{w})u,$$

$$\frac{dM_y}{dt} = \oiint \rho dA w \cos (\mathbf{n}, \mathbf{w})v,$$

$$\frac{dM_z}{dt} = \oiint \rho dA w \cos (\mathbf{n}, \mathbf{w})w,$$

where $\mathbf{n}$ is the normal (considered positive outward). If dA is perpendicular to the x-direction, the displaced fluid is $\rho dA u$ and

$$\frac{dM_x}{dt} = \int \int \rho dA u^2.$$

When dA is perpendicular to the y-axis, the displaced fluid is

$$\frac{dM_x}{dt} = \int \int \rho dA uv,$$

and so on.

We have therefore found that in steady flow the rate of change of momentum is equal to the flow of momentum across the fixed boundary surface.

The right-hand side of Eq. (1), being the sum of all external forces, in a non-viscous fluid is composed of:

1. Pressures on the bounding surface: $-\oiint p d\mathbf{A}$ [vectorial integration; x-component $-\oiint p d\mathbf{A} \cos (\mathbf{n}, x)$, etc.]. The negative sign occurs because the pressure is directed inward and the positive direction of the normal is outward.

2. Impressed forces, particularly gravity $\int\int\int \rho \mathbf{g} dV$.

3. Other forces, such as external forces exerted on bodies in the fluid, *i.e.*, $\Sigma\mathbf{P}_e$. In the liquid there may be rigid bodies on which external forces are acting. We can of course consider the surfaces of these bodies as boundaries of the fluid, and the pressures they exert can be included in the surface integral of the pressure p. It is however much simpler to deal with these additional forces directly.

The equation of momentum therefore is

$$\oiint \rho d\mathbf{A}\circ\mathbf{w}\mathbf{w} = \int\int\int \rho \mathbf{g} dV - \oiint p d\mathbf{A} + \Sigma\mathbf{P}_e \tag{3}$$

or, if gravity can be neglected as is true in many cases,

$$\oiint \rho d\mathbf{A}\circ\mathbf{w}\mathbf{w} + \oiint p d\mathbf{A} = \Sigma\mathbf{P}_e. \tag{3a}$$

In words: When the motion of a non-viscous fluid is steady, the flow of momentum through a fixed surface bounding a definite volume of the fluid is equal to the resultant of the pressure integral over this surface and the body forces and other forces acting on the enclosed fluid.

Fig. 170.—The resultant of all pressure forces exerted by the rigid guiding surface is equal to the momentum carried through the surfaces A_1 and A_2.

In many cases we are not so much interested in the outside forces exerted on the fluid, but conversely we want the forces exerted by the fluid on bodies in it (and not belonging to the system) such as turbine blades or airfoils. This question can be answered by changing the sign of the corresponding forces in the previous equation. The relation is analogous to the difference between action and reaction. The analogy can be seen more clearly by considering the flow between two curved walls (Fig. 170). The broken line is the bounding surface. In this case there are no obstacles in the fluid and thus by neglecting gravity the equation of momentum is

$$\oiint \rho d\mathbf{A} \circ \mathbf{w} \mathbf{w} = -\oiint p d\mathbf{A}.$$

Since the walls are rigid there can only be flow of momentum through the surfaces $\mathbf{A_1}$ and $\mathbf{A_2}$. The external normal at the surface $\mathbf{A_1}$ is opposite to the direction of flow so that the momentum entering through this surface is

$$-\rho A_1 |\mathbf{w_1}| \mathbf{w_1}.$$

The force $\mathbf{P_1}$ corresponding to this flow of momentum is equal to $\rho A_1 w_1^2$ and in a direction opposite to that of $\mathbf{w_1}$. $\mathbf{P_2}$ is equal to $\rho A_2 w_2^2$ and is in the direction of $\mathbf{w_2}$ since the normal at $\mathbf{A_2}$ is parallel to and in the same direction as the velocity.

The resultant of $\mathbf{P_1}$ and $\mathbf{P_2}$, *i.e.*, $\mathbf{P}$ corresponds to the total flow of momentum through $\mathbf{A_1}$ and $\mathbf{A_2}$, and is the resultant of all pressures exerted by the walls on the fluid.

By the principle of action and reaction there must be a force equal and opposite to $\mathbf{P}$ exerted by the fluid on the walls. The forces $\mathbf{R_1}$ and $\mathbf{R_2}$ which are reactions corresponding to $\mathbf{P_1}$ and $\mathbf{P_2}$ are equal and opposite to their respective counterparts (Fig. 171). Therefore we can say that the entering momentum corresponds to a force in an opposed direction and the momentum that

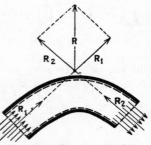

Fig. 171.—By the principle of action and reaction, the forces opposed to those of Fig. 170 produce a force $\mathbf{R}$ on the rigid wall due to the action of the fluid.

is leaving corresponds to a force in the same direction (Fig. 170) when we are considering the effect of external forces *on* the system. If on the other hand we are considering the forces exerted *by* the fluid on external objects, the directions of the above forces must be reversed. The entering momentum corresponds to a force in its direction and the emerging momentum to a force in a direction opposed to its own (Fig. 171). In the first case the force is always directed out of the liquid and in the second always into it.

In many cases, as will be seen from examples, only certain components of the equation of momentum are used, the equation of course being in vector form. We also usually choose the surface of integration so that there is no flow of momentum through some part of it.

The momentum equation, (3), can be derived quite easily also from Euler's equation. On page 110 we had

$$\rho\left(\frac{\partial \mathbf{w}}{\partial t} + \mathbf{w}\circ\nabla\mathbf{w}\right) = \rho\mathbf{g} - \text{grad } p.$$

If we make use of the relation

$$\nabla\circ(\mathbf{ww}) = (\nabla\circ\mathbf{w})\mathbf{w} + \mathbf{w}\circ\nabla\mathbf{w}$$

and assume an incompressible fluid

$$(\text{div } \mathbf{w} = \nabla\circ\mathbf{w} = 0),$$

we obtain on multiplying by dV and integrating over the whole of the bounded volume

$$\iiint \rho\frac{\partial \mathbf{w}}{\partial t}dV + \iiint \rho\nabla\circ(\mathbf{ww})dV =$$
$$\iiint \rho\mathbf{g}dV - \iiint \text{grad } pdV.$$

But Gauss's theorem (see Art. 46) gives

$$\iiint \nabla\circ(\mathbf{ww})dV = \oiint d\mathbf{A}\circ\mathbf{ww}$$

and

$$\iiint \text{grad } pdV = \oiint pd\mathbf{A}.$$

If now we assume further that the flow is steady $(\partial\mathbf{w}/\partial t = 0)$, we get

$$\oiint \rho d\mathbf{A}\circ\mathbf{ww} = \iiint \rho\mathbf{g}dV - \oiint pd\mathbf{A}$$

which agrees with Eq. (3).

In the above equation the forces $\Sigma\mathbf{P}_e$ are included in the pressure integral over the boundary of the liquid, the surfaces of obstacles being considered as boundaries.

Applying now the fundamental theorem of the mechanics of discrete particles, stating that the rate of change of the moment of momentum round a point or axis is equal to the moment of the external forces,

$$\frac{d}{dt}\left(\sum m\mathbf{r} \times \mathbf{w}\right) = \sum\mathbf{r} \times \mathbf{P},$$

we get a theorem of moment of momentum analogous to the relation between angular momentum and force

$$\oiint \rho d\mathbf{A}\circ\mathbf{w}\,\mathbf{r} \times \mathbf{w} = \iiint \rho\mathbf{r} \times \mathbf{g}dV - \oiint p\mathbf{r} \times d\mathbf{A} + \Sigma\mathbf{r} \times \mathbf{P}_e.$$

101. Extension of the Momentum Theorem to Fluid Motion with a Steady Mean Flow.—Equations of momentum can be formed for non-steady flow but in general nothing can be derived from them, for the integral $\iiint \rho(\partial \mathbf{w}/\partial t)dV$ usually cannot be transformed into a surface integral. The advantage of the momentum theorem of needing only a knowledge about the conditions on the bounding surface is thus lost. The application to non-steady flow however is possible when dealing with periodic motions or with motions having a steady mean value of the velocity, *i.e.*, consisting of a "basic motion" and a superposed regular or irregular variation. The momentum theorem yields results for the mean values over long periods of time during which the irregular variations cancel.

If the velocity of a fluid particle in the region under consideration is

$$\mathbf{w} = \overline{\mathbf{w}} + \mathbf{w}',$$

where $\overline{\mathbf{w}}$ is the steady mean and $\mathbf{w}'$ the variation from the mean, the mean value of $\mathbf{w}$ (the mean value is always indicated by a bar over the symbol) is

$$\overline{\mathbf{w}} = \overline{\mathbf{w}} + \overline{\mathbf{w}'},$$

giving $\overline{\mathbf{w}'} = 0$. This means that the mean value of the variation from the mean value of velocity is equal to zero by definition. If in applying the momentum theorem to the interior of the fluid we replace the momentum at each point by its mean value, the rate of change of this momentum is zero, since the mean value by definition has no variation.

On the bounding surface, however, we have to form the mean values of second-degree terms such as u^2, uv, etc., where u, v, w are Cartesian components of $\mathbf{w}$. But

$$\overline{u^2} = \overline{(\bar{u} + u')^2} = \bar{u}^2 + 2\overline{\bar{u}u'} + \overline{u'^2}$$

and, since $\overline{u'} = 0$,

$$\overline{u^2} = \bar{u}^2 + \overline{u'^2}$$

Similarly,

$$\overline{uv} = \overline{(\bar{u} + u')(\bar{v} + v')} = \bar{u}\bar{v} + \overline{\bar{u}'v} + \overline{\bar{u}v'} + \overline{u'v'}$$

and, since $\overline{u'} = \overline{v'} = 0$,

$$\overline{uv} = \bar{u}\bar{v} + \overline{u'v'}.$$

The mean value of u'^2 is always positive, and that of $\overline{u'v'}$ can be positive, zero, or negative. It is positive for example when u' and v' are both positive or both negative.

If $p = \bar{p} + p'$ is the pressure, where $\bar{p}$ is the steady mean value and p' the deviation from the mean, then as with the velocity, the mean value of p' is zero.

The momentum theorem is also applicable when there are no steady mean values, but the flow is partially steady and periodicities appear in part of the fluid (as in the Kármán vortex trail. In this case the flow pattern round the obstacle is the same after each period. The frame of reference is so chosen that the vortex system is steady. The bounding surface consists of a plane through the vortex system and of a surface surrounding the obstacle and lying completely in the undisturbed fluid (Fig. 172). If we compare the positions before and after an interval of a period, a volume integral due to

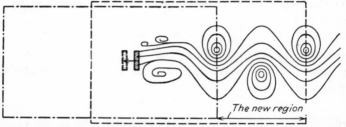

The new region

Fig. 172.—Application of the theorem of momentum to the periodic Kármán trail of vortices.

the increase of momentum in the interior of the fluid must be added to the pressure and momentum integrals over the bounding surface. The equation of momentum in this case gives the mean value of the force over a period.

102. Applications of the Theorem of Momentum.—A few simple examples will show how quickly and simply the application of the theorem of momentum yields general results about the flow without having to know anything about the details of the motion.

Case 1. *Efflux from an Open Vessel.*—We neglect gravity in the issuing jet, and, since there are no foreign bodies in the fluid experiencing external forces, Eq. (3a) gives

$$\oiint \rho\, d\mathbf{A} \circ \mathbf{w}\mathbf{w} = -\oiint p\, d\mathbf{A} = \mathbf{P}.$$

The surface of integration is as shown in Fig. 173, and we assume that the velocity **w** over the cross section of the jet **A** is constant. Since ρ is constant, the flow of momentum is

$$\oiint \rho d\mathbf{A} \circ \mathbf{w}\mathbf{w} = \rho \mathbf{A}|\mathbf{w}|\mathbf{w} = \rho A w^2 \frac{\mathbf{w}}{w}.$$

The term $\mathbf{w}/w$ is a directional factor whose numerical value is unity.

The impulse flow issuing from the vessel is equivalent to an equal force in the opposite direction by Art. 100. Let the pressure inside the vessel be denoted by p_1 and the external atmospheric pressure (which is also assumed to act in the jet) by p_0. Bernoulli's equation of page 118 gives

$$w^2 = \frac{2(p_1 - p_0)}{\rho}$$

so that the reaction force is

$$P = 2A(p_1 - p_0).$$

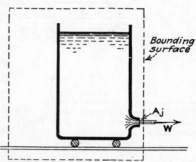

If the vessel stands on frictionless wheels, it experiences an acceleration produced by the force P in a direction opposed to that of the issuing jet, and if the vessel is to be kept at rest, it must be subjected to a force R which is equal to $-P$.

Fig. 173.—The fluid jet to the right gives rise to a reaction force to the left, which would produce an acceleration of the open vessel to the left—if there were no friction between the vessel and the surface on which it stands.

The reason for the appearance of the factor 2 in the above expression is that in addition to the diminution of static pressure $A(p_1 - p_0)$ at the opening on account of the flow toward it, there is an additional fall of pressure along the walls in the neighborhood of the opening. The theorem of momentum shows that the force corresponding to this fall in pressure is exactly equal to that due to the difference of pressure over the cross section of the jet.

Case 2. *The Contraction Coefficient of the Borda Mouthpiece.*— Let A be the cross section of the mouthpiece and A_j the area of the bounding surface pierced by the emerging jet, so that the coefficient of contraction α is defined by $\alpha = A_j/A$. The bounding surface is shown in Fig. 174, and by making assumptions identical with those of the preceding example we get

$$\oiint p d\mathbf{A} = \rho \oiint d\mathbf{A} \circ \mathbf{w}\mathbf{w} = \rho A_j w^2 \frac{\mathbf{w}}{w} = 2A_j(p_1 - p_0)\frac{\mathbf{w}}{w}.$$

But the difference of pressure over all surfaces which have a component of pressure in the direction of the jet is $A(p_1 - p_0)$, so that the loss over the area of the mouthpiece, $A(p_1 - p_0)$,

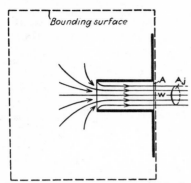

must be equal to the momentum of the jet. Hence

$$A(p_1 - p_0) = 2A_j(p_1 - p_0)$$

or

$$A_j = \tfrac{1}{2}A$$

or

$$\alpha = \tfrac{1}{2}.$$

Case 3. The Reaction in a Curved Channel Caused by the Flow of Fluid.—Referring to Art. 100, Fig. 171, it was seen that the flow of momentum through

Fig. 174.—Determination of the A_1 and A_2, coefficient of contraction of a Borda mouthpiece.

$$\rho A_1 w_1{}^2 \frac{\mathbf{w_1}}{w_1} - \rho A_2 w_2{}^2 \frac{\mathbf{w_2}}{w_2},$$

corresponds to a reaction $\mathbf{R} = \mathbf{R}_1 + \mathbf{R}_2$, exerted by the fluid on the bounding surface. If we neglect the effect of gravity, this force is absorbed by the pressure of the fluid on the bounding surface. The surface pressure however is composed only of the forces exerted by the channel walls and those on the end surfaces, namely, $\mathbf{A}_1 p_1 + \mathbf{A}_2 p_2$. The reaction pressure of the fluid on the walls ($\mathbf{R}_w$) therefore is:

$$\mathbf{R}_w = \rho A_1 w_1{}^2 \frac{\mathbf{w_1}}{w_1} - \rho A_2 w_2{}^2 \frac{\mathbf{w_2}}{w_2}$$
$$- \mathbf{A}_1 p_1 - \mathbf{A}_2 p_2.$$

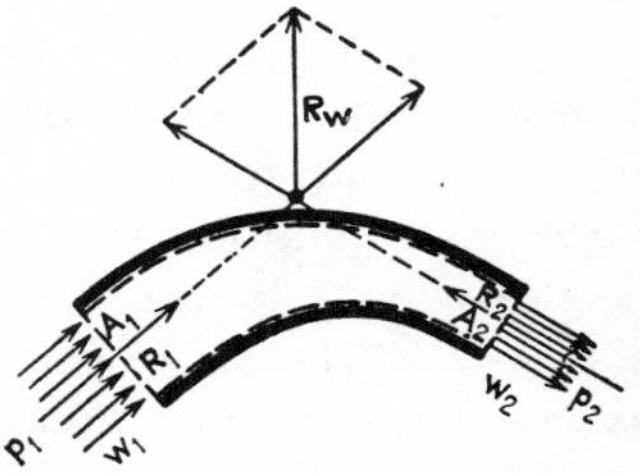

When the surfaces $\mathbf{A}_1$ and $\mathbf{A}_2$ are at right angles to the direction of flow across them, then, since the vector $\mathbf{A}$

Fig. 175.—Reaction of fluid flowing through curved channel or pipe.

is opposed in direction to $\mathbf{w}_1$ and since $\mathbf{A}_2$ is the same direction as $\mathbf{w}_2$, we can also write:

$$\mathbf{R}_w = -\rho\mathbf{A}_1 w_1{}^2 - \mathbf{A}_1 p_1 - \rho\mathbf{A}_2 w_2{}^2 - \mathbf{A}_2 p_2$$

or

$$\mathbf{R}_w = -\mathbf{A}_1(\rho w_1{}^2 + p_1) - \mathbf{A}_2(\rho w_2{}^2 + p_2).$$

In Fig. 175 the resultant reaction of the fluid on the walls is built up from its components: $-\mathbf{A}_1(\rho w_1{}^2 + p_1)$ and $-\mathbf{A}_2(\rho w_2{}^2 + p_2)$.

Case 4. *Sudden Change in Section of a Tube.*—A fluid jet emerges from a cylindrical tube of cross section A_1 into a section A_2 (Fig. 176). Since the emerging jet of velocity w_1 forms a surface of separation in the wider tube and thus is in a position of unstable equilibrium, it mixes with the surrounding fluid and at a sufficient distance to the right the irregularities have died down and there again exists a uniform flow with a mean velocity w_2. Let the pressure in the jet when leaving the narrow tube be p_1, and the pressure in the wider one after complete mixing be p_2. The pressure difference $p_2 - p_1$ can be calculated by applying the momentum theorem without knowing anything about the details of the mixing process.

We neglect gravity and assume that the fluid is homogeneous, and, since there are no foreign bodies in the fluid, we have

$$\rho \oiint d\mathbf{A} \circ \mathbf{w}\mathbf{w} = -\oiint p d\mathbf{A}.$$

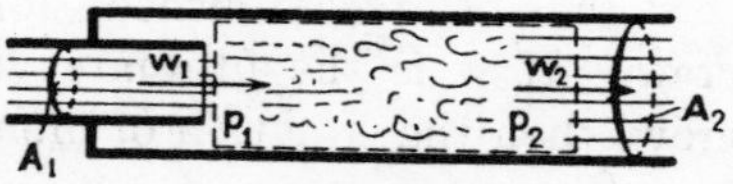

FIG. 176.—Sudden widening of a pipe.

The surface of integration is as shown in Fig. 176 and the only flow of momentum is through the surfaces A_1 and A_2. Hence

$$\rho A_2 w_2{}^2 - \rho A_1 w_1{}^2 = -A_2(p_2 - p_1)$$

and, making use of the equation of continuity of an incompressible fluid in the form $A_1 w_1 = A_2 w_2$, we get

$$p_2 - p_1 = \rho w_2(w_1 - w_2),$$

which is the increase in pressure.

If the area of the tube increases very gradually from A_1 (and pressure p_1) to the larger area A_2 (and corresponding pressure p_2') Bernoulli's equation can be applied, giving for the increase of pressure due to the loss in velocity

$$p_2' - p_1 = \frac{\rho}{2}(w_1{}^2 - w_2{}^2).$$

The difference between the final pressures for gradual and sudden widening is thus

$$p_2' - p_2 = p_2' - p_1 - (p_2 - p_1) = \frac{\rho}{2}(w_1{}^2 - w_2{}^2) - \rho w_2(w_1 - w_2)$$

$$= \frac{\rho}{2}(w_1 - w_2)^2,$$

and we realize that the final pressure in the case of sudden widening is smaller than if the change took place gradually. A diminu-

tion of velocity can be obtained by a *sudden* increase in cross section only at the expense of a certain loss of head. The greater the drop in velocity desired, the greater is the loss of head incurred.

The last equation determining the loss of pressure in a gradually widening tube has been called the "Carnot impact formula" on account of its similarity to the formula giving the loss of energy when two inelastic bodies impinge, although "impact" does not enter into this problem at all. In both these cases however there is an irreversible diminution of velocity.

Case 5. *Sustaining Heavy Bodies in Fluids.*—In order to make bodies (heavier than air) rise against the action of gravity and to keep them in that position, there must be a continuous downward acceleration of fresh masses of air beneath the body. This can be accomplished, for example, by some helicopter screw. If **w** is the ultimate velocity of the air seized by the screw, then the equation of momentum is

$$\oiint \rho d\mathbf{A} \circ \mathbf{w}\mathbf{w} = -\oiint p d\mathbf{A} + \Sigma \mathbf{P}_e.$$

If a very large bounding surface is taken so that the pressure differences in the air that has been pushed downward can be neglected, the reaction on the screw propeller is a force $\rho A w^2$ upward, when A is the area of the stream that is accelerated downward. If M is the mass of air that is set in motion per unit time, then the reaction force is

$$\mathbf{P} = -M\mathbf{w}.$$

When the screw does no other work except keeping itself in the air, the resultant work done per unit time W is equal to the energy per unit time of the downward stream,

$$W = M\frac{w^2}{2} = P\frac{w}{2}.$$

We therefore see that the application of the momentum theorem to an air mass shows that the body can be kept in the air, either by imparting a large velocity to a small mass, or by accelerating a large air mass to a small velocity. However the power to be expended on the helicopter screw becomes smaller if the downward velocity is kept small and if as much air as possible is set into motion.

Case 6. *Two-dimensional Jet against an Inclined Plate.*—There are two points of interest here: the force exerted by the jet

on the plate, and the way in which the jet is diverted up and down the plate. In Fig. 177, a is the width of the jet and **w** its velocity, a_1 and a_2 are the widths of the partial jets up and down the plate. Let p_0 be the pressure in the advancing jet, O the point of intersection of the center line of the jet and the plate, and e the distance of the center of pressure from O.

Since the pressure in those parts of the partial jets, where the flow is already parallel to the plate, is that of the atmosphere p_0, it follows that the velocities in these parts have a value w. The surface of integration is shown in Fig. 177, and neglecting gravity we have

$$\rho \oiint d\mathbf{A} \circ \mathbf{w}\mathbf{w} = -\oiint pd\mathbf{A} = \mathbf{P}.$$

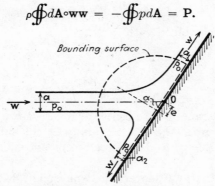

Fig. 177.—Two-dimensional flow of a water jet against an inclined plate.

Let us decompose this vector equation into its component equations perpendicular and parallel to the plate, and, assuming that the depth of the system perpendicular to the plane of the diagram is unity, we get:

component perpendicular to plate,

$$P = \rho a w^2 \sin \alpha;$$

component parallel to plate,

$$0 = \rho a_1 w^2 - \rho a_2 w^2 - \rho a w^2 \cos \alpha$$

or

$$a_1 - a_2 = a \cos \alpha.$$

From the principle of continuity we have

$$a_1 + a_2 = a$$

so that

$$a_1 = a\frac{1 + \cos \alpha}{2}$$

and

$$a_2 = a\frac{1 - \cos \alpha}{2}.$$

We have now answered the questions relating to the pressure on the plate and the distribution of the jet, but the position of the center of pressure on the plate is also of interest. In order to determine this distance e we consider the moment of momentum about O. Since the distances between the central lines of flow of the three jets and the axis at O, round which the moment of momentum is taken, are 0, $a_1/2$, and $a_2/2$, we have

$$\rho a_1 w^2 \frac{a_1}{2} - \rho a_2 w^2 \frac{a_2}{2} = Pe$$

which, when combined with

$$\rho a w^2 \sin \alpha = P,$$

gives

$$e = \frac{1}{\sin \alpha}\left(\frac{a_1{}^2 - a_2{}^2}{2a}\right) = \frac{a}{2} \cotg \alpha.$$

This is the distance of the center of pressure from O.

The momentum theorem naturally is not sufficiently fundamental to enable us to determine the pressure distribution along the plate.

103. The Energy Theorem for Non-steady Motion of Incompressible Fluids.—In much the same way as we derived a theorem for the transport of momentum in Art. 100, we can form a corresponding theorem for the transport of energy.

Our starting point is the general fact that the increase of energy in a system is equal to the work done on the system by external forces, or the rate of change of energy is equal to the work done on the system.

In our calculations all kinds of energy must be taken into account, such as kinetic, potential, elastic, and thermal, so that when considerable amounts of work are done by friction and energy is thus dissipated, our considerations are of little value. We cannot calculate the dissipation without a knowledge of the details of the motion in the interior of the fluid, but the energy theorem (as well as the momentum theorem) is only useful when it enables us to calculate resultant forces from a knowledge of the boundary conditions. Hence energy considerations can only be usefully applied when the amount of frictional dissipation is negligible.

The results of the energy theorem in steady flow when friction is neglected are always trivial, for the work done on the obstacle at rest is zero, and in the fluid the energy is constant so that the application of the theorem of energy regularly yields results of the form zero equals zero.

If w_n is the normal component of velocity of a fluid particle on the bounding surface, considered positive if in the outward direction, the mass sent through an element dA of the bounding surface per unit time is

$$dm = dA\rho w_n,$$

and the corresponding kinetic energy is

$$dE = dA\rho w_n \frac{w^2}{2}.$$

On the other hand the action of pressure in displacing the fluid surface is to be taken into account so that the work done per unit time is

$$dW = -p \, dA \, w_n.$$

In the steady case the energy theorem is

$$\oiint dE = \oiint dW$$

or

$$0 = \oiint \rho w_n \left(\frac{w^2}{2} + \frac{p}{\rho} \right) dA.$$

The bracketed expression in the integrand however is constant for irrotational flow of an incompressible fluid and can therefore be put outside the sign of integration. The remaining integral simply tells us that the total flow of matter through the closed bounding surface is equal to zero (sources and sinks have been excluded), so that the theorem gives no information whatever.

If U is the thermal energy per unit mass (including any possible elastic energy), measured in mechanical units, the convective transport of energy per unit time through the element dA is

$$\rho \, dA \, w_n \left(\frac{w^2}{2} + U \right).$$

On the other hand the work done by the pressure is

$$-p \, dA \, w_n.$$

Combining these two expressions, we get

$$\rho \, dA \, w_n \left(\frac{w^2}{2} + U + \frac{p}{\rho} \right),$$

or writing $v = 1/\rho$ for the volume of unit mass,

$$\rho \, dA \, w_n \left(\frac{w^2}{2} + U + pv \right).$$

If we assume an adiabatic change of state, then

$$U + \int p \, dv = \text{const.}$$

and Bernoulli's equation gives

$$\frac{w^2}{2} + \int \frac{dp}{\rho} = \frac{w^2}{2} + \int v \, dp = \text{const.}$$

Hence

$$\frac{w^2}{2} + U + \int (p \, dv + v \, dp) = \frac{w^2}{2} + U + pv = \text{const.},$$

which is identical with the expression above. Along a stream tube $\rho \, dA \, w_n$ is constant so that the equation of energy again tells us nothing new.

When the flow is not steady, the energy theorem can be of some value as for example when energy is added to the system by the non-steady motion of a body like a propeller. If we assume in this case that the propeller causes an increase of pressure in the fluid, then this pressure will be converted into additional kinetic energy of the fluid, which must be equal to the work done on the propeller (the coordinate system is chosen so that the propeller is at rest). For another coordinate system, with respect to which the undisturbed fluid is at rest, the kinetic energy per unit time is smaller and equal to the difference between the work done by the motor on the propeller and that done by the propeller thrust on the motion.

A different way of applying the energy theorem is useful in cases where an obstacle generates a regular eddy motion in a fluid that was originally at rest, as is the case in an airplane wing or propeller. The system of reference is usually chosen to be at rest relative to the undisturbed fluid. The obstacle, *e.g.* the propeller, moves forward along its axis with a velocity V and rotates with a constant angular velocity ω. If T is the torque of the propeller shaft and S the thrust, then the work done

by the propeller is $T\omega - SV$, and the energy equilibrium is expressed by

$$\frac{dE}{dt} + \oiint p \, dA \, w_n = T\omega - SV,$$

where E is the kinetic energy of the fluid and the integral is the work done by the pressure on the bounding surface. For this surface we conveniently take a plane perpendicular to the direction of the velocity of advance behind the propeller combined with some other surface (as in Fig. 178) enclosing the propeller and passing entirely through the undisturbed fluid.

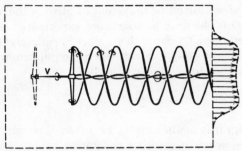

Fig. 178.—Vortex system due to a propeller moving in the direction of its axis.

Since the energy content per unit length of the bounded fluid mass is constant at a sufficient distance behind the propeller, the rate of change of kinetic energy dE/dt is equal to the flow of energy through the rear plane of the bounding surface perpendicular to V plus the kinetic energy between this rear plane (which is at rest) and another plane A moving with velocity V to the left and coinciding with the rear plane at the time $t = 0$. Expressed mathematically this is

$$\frac{dE}{dt} = \oiint w_n \rho \frac{w^2}{2} dA + V \oiint \rho \frac{w^2}{2} dA = \oiint \frac{\rho w^2}{2}(V + w_n) dA,$$

where the integration extends over the bounding surface.

This takes care of the first term in the energy-equilibrium equation. Consider next the second term $\oiint p \, dA \, w_n$. Although the pressures decrease proportional to the square of the distance, the elements of area dA increase at the same rate, and, since w_n decreases to zero outside the propeller stream, the integral converges to zero when the closing plane of the bounding surface

is far behind the propeller.[1] On the other hand the pressure integral generally does not vanish in the vortex region. Thus the energy equilibrium becomes

$$T\omega - SV = V\iint \frac{\rho w^2}{2} dA + \iint w_n \left(\rho \frac{w^2}{2} + p\right) dA,$$

where both integrals need only be taken in the vortex region and its immediate neighborhood behind the propeller. If the propeller has a small load, *i.e.*, does little work, w_n is everywhere small in comparison with V, and as a first approximation the second integral can be neglected in comparison with the first.

In the case of an airfoil or a system of airfoils the work done is DV, where D is the drag or resistance experienced in the fluid (supposed frictionless). Since here w_n is negligibly small, the last integral in the above given energy-equilibrium equation vanishes so that we have approximately,

$$D = \iint \frac{\rho w^2}{2} dA,$$

where the integral is again only to be taken through the vortex system. Thus, when neglecting the terms proportional to w_n, the resistance is equal to the kinetic energy per unit length that is left behind.

---

[1] Great care has to be exercised when applying such limiting processes to the momentum theorem. The momentum or pressure integrals over the part of the bounding surface in the undisturbed fluid are often different from zero.

CHAPTER XV

THE EQUATION OF NAVIER-STOKES FOR VISCOUS FLUIDS

104. The Fundamental Equation of Fluid Mechanics.—
Newton's fundamental law of mechanics,

Mass × acceleration = force,

applied to the unit of volume of a fluid, is

$$\rho \frac{D\mathbf{w}}{dt} = \mathbf{F} + \mathbf{G}, \tag{1}$$

where the total force has been decomposed in body forces $\mathbf{F}$ and
surface forces $\mathbf{G}$. Leaving out of the discussion systems in which
centrifugal forces, Coriolis forces, etc., occur, the only body force
is the force of gravity per unit volume

$$\mathbf{F} = \rho \mathbf{g}.$$

In case friction forces are neglected, the only surface force per
unit volume is the pressure drop, grad p (see Art. 56). In the
present chapter the assumption
of negligible viscosity will be
dropped, and it will be investigated how the motion of an
actual fluid differs from that of
the ideal fluid without friction.

It is assumed that $\mathbf{G}$ is an
analytic function of space and
that the fluid under consideration is an isotropic body having no preferred directions. In
particular the coefficients of

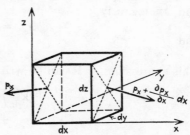

Fig. 179.—The forces acting on the
dy-dz sides of an element of viscous
fluid.

viscosity are assumed to be independent of the direction. Considering the infinitesimal cube of Fig. 179 with volume
$dV = dxdydz$, the total surface force consists of the three vectors:

$$\frac{\partial \mathbf{p}_x}{\partial x}dxdydz, \qquad \frac{\partial \mathbf{p}_y}{\partial y}dydzdx, \qquad \frac{\partial \mathbf{p}_z}{\partial z}dzdxdy.$$

In order to use the same notations as in the theory of elasticity, tensile forces are taken as being positive and compressive force as negative. Therefore the resultant force becomes

$$\mathbf{R} = \left(\frac{\partial \mathbf{p}_x}{\partial x} + \frac{\partial \mathbf{p}_y}{\partial y} + \frac{\partial \mathbf{p}_z}{\partial z}\right)dV.$$

105. Decomposition of the Total Surface Force into the Elements of a Stress Tensor.—Decomposing the vectors $\mathbf{p}_x$, $\mathbf{p}_y$, $\mathbf{p}_z$ into their rectangular components, *e.g.*, $\mathbf{p}_x = \mathbf{i}\sigma_x + \mathbf{j}\tau_{xy} + \mathbf{k}\tau_{xz}$, the total force per unit volume can be written as follows:

$$\mathbf{R} = \mathbf{i}\left(\frac{\partial \sigma_x}{\partial x} + \frac{\partial \tau_{xy}}{\partial y} + \frac{\partial \tau_{xz}}{\partial z}\right) \quad (x\text{-component}),$$

$$\mathbf{j}\left(\frac{\partial \tau_{xy}}{\partial x} + \frac{\partial \sigma_y}{\partial y} + \frac{\partial \tau_{yz}}{\partial z}\right) \quad (y\text{-component}),$$

$$\mathbf{k}\left(\underbrace{\frac{\partial \tau_{xz}}{\partial x}}_{yz\text{-plane}} + \underbrace{\frac{\partial \tau_{yz}}{\partial y}}_{zx\text{-plane}} + \underbrace{\frac{\partial \sigma_z}{\partial z}}_{xy\text{-plane}}\right) \quad (z\text{-component}).$$

In the derivation of this expression, the fact has been used that the moment of the forces about an arbitrary axis must be zero,[1] for instance,

$$\tau_{xy}dydzdx = \tau_{yx}dxdzdy$$

so that

$$\tau_{xy} = \tau_{yx}$$

and, analogously,

$$\tau_{xz} = \tau_{zx}$$

and

$$\tau_{yz} = \tau_{zy}.$$

This expression for $\mathbf{R}$ can be written in a different form:

$$\mathbf{R} = \left(\mathbf{i}\frac{\partial}{\partial x} + \mathbf{j}\frac{\partial}{\partial y} + \mathbf{k}\frac{\partial}{\partial z}\right)\circ(\mathbf{ii}\sigma_x + \mathbf{ij}\tau_{xy} + \mathbf{ik}\tau_{xz}$$
$$+\mathbf{ji}\tau_{xy} + \mathbf{jj}\sigma_y + \mathbf{jk}\tau_{yz}$$
$$+\mathbf{ki}\tau_{xz} + \mathbf{kj}\tau_{yz} + \mathbf{kk}\sigma_z),$$

[1] In writing down this relation, the fluid cube has been tacitly replaced by an elastic cube. This is correct since it is generally assumed that the stress condition in a viscous fluid is of the same kind as the stress condition in an elastic body with the only difference that the stresses in the elastic body are proportional to the deformations, whereas in a fluid they are proportional to the rate of change of deformation. For a cube of fluid the above equation cannot be found by referring to the case of equilibrium, since the definition of a fluid is that in the case of equilibrium all shear stresses disappear.

or introducing the operator

$$\nabla = \mathbf{i}\frac{\partial}{\partial x} + \mathbf{j}\frac{\partial}{\partial y} + \mathbf{k}\frac{\partial}{\partial z},$$

this becomes

$$\mathbf{R} = \nabla \circ \mathbf{i} \begin{vmatrix} \mathbf{i} & \mathbf{j} & \mathbf{k} \\ \sigma_x & \tau_{xy} & \tau_{xz} \\ \tau_{xy} & \sigma_y & \tau_{yz} \\ \tau_{xz} & \tau_{yz} & \sigma_z. \end{vmatrix}$$

Or, written in a shorthand manner,

$$\mathbf{R} = \nabla \circ \Pi, \tag{2}$$

where

$$\Pi = \begin{Bmatrix} \sigma_x & \tau_{xy} & \tau_{xz} \\ \tau_{xy} & \sigma_y & \tau_{yz} \\ \tau_{xz} & \tau_{yz} & \sigma_z \end{Bmatrix}$$

is an expression characteristic for the state of stress in the elementary volume under consideration. This so-called "stress

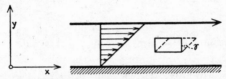

Fig. 180.—Velocity distribution in a viscous fluid between a moving and a stationary plate.

tensor" is a symmetrical tensor on account of the relations $\tau_{xy} = \tau_{yx}$, etc., and consequently is determined by six quantities. Its relation to the total surface force is expressed by Eq. (2).

106. Relation of the Elements of the Stress Tensor to the Corresponding Rates of Change of Deformation.—After having decomposed the total surface force **G** into its constituent elements (appearing in the tensor Π), we now proceed to find the relation between the state of stress Π and the rate of change of deformation **w**. To this end we first find a relation between the individual elements of the stress tensor and the corresponding rates of change of deformation.

For the case of steady flow, pictured in Fig. 180, we know from experiment that the shear force per unit surface of the upper moving plate has the following relation to the velocity gradient:

$$\tau = \mu\frac{\partial u}{\partial y},$$

where μ, the viscosity, not only is different for various fluids but varies greatly with the temperature for any given fluid. Denoting by γ the amount by which an originally right angle in the fluid element has diminished, we have

$$\tau = \mu \frac{\partial \gamma}{\partial t}.$$

In other words, the shear stress is proportional to the rate of change of deformation, the proportionality factor being μ.

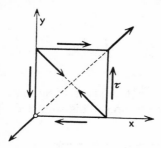

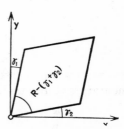

Fig. 181.—Shear stresses on an infinitesimal cube.

Fig. 182.—Change in the right angle due to shear stresses.

Considering in Fig. 181 an infinitesimal square, the shear stresses $\tau_{xy} = \tau_{yx} = \tau$ cause a certain change per second in the original right angle (Fig. 182). Therefore

$$\tau = \mu \frac{\partial \gamma}{\partial t} = \mu\left(\frac{\partial \gamma_1}{\partial t} + \frac{\partial \gamma_2}{\partial t}\right) = \mu\left(\frac{\partial u}{\partial y} + \frac{\partial v}{\partial x}\right).$$

Applying the same reasoning to τ_{yz} and τ_{xz}, we get

$$\tau_{xy} = \mu\left(\frac{\partial u}{\partial y} + \frac{\partial v}{\partial x}\right),$$

$$\tau_{yz} = \mu\left(\frac{\partial v}{\partial z} + \frac{\partial w}{\partial y}\right),$$

and

$$\tau_{xz} = \mu\left(\frac{\partial w}{\partial x} + \frac{\partial u}{\partial z}\right).$$

Now we turn our attention to the normal stresses σ_x, σ_y, σ_z. In the square of Fig. 183, we draw in thin lines another square standing on one of its points. Under the influence of the shear stresses this square is deformed into a rectangle as indicated. The equilibrium of forces in a triangle cut off from the main

square by means of a diagonal shows that the stresses in the diagonals are either tensile or compressive. From Fig. 184 we see that

$$2a\tau \sin \frac{\pi}{4} - a\sqrt{2}\sigma_1 = 0$$

or, since

$$\sin \frac{\pi}{4} = \frac{1}{\sqrt{2}},$$

we have

$$\sigma_1 = \tau.$$

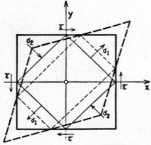

Fig. 183.—Deformation of the square standing on its point into a rectangle due to the shear stresses τ.

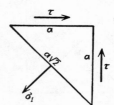

Fig. 184.—Equilibrium between normal stresses and shear stresses.

In the same manner we find from the equilibrium of a triangle cut off by the other diagonal that

$$\sigma_2 = -\tau$$

and therefore

$$\sigma_1 - \sigma_2 = 2\tau.$$

Now we determine how much the diagonal of the square has changed its length when it is deformed according to Fig. 185. In this case, $ABCD$ transforms into $AB'C'D'$, and the right angle at A becomes $90° - \gamma$. Since

$$\overline{DD'} = \overline{EM'} = \overline{AB} \cdot \frac{\gamma}{2},$$

and consequently

$$\overline{MM'} = \frac{\overline{EM'}}{\sqrt{2}} = \frac{\overline{AB}}{\sqrt{2}} \cdot \frac{\gamma}{2},$$

and since

$$\overline{AM} = \frac{\overline{AB}\sqrt{2}}{2},$$

we find for the elongation ϵ_1 of the diagonal $\overline{AC}$

$$\epsilon_1 = \frac{\overline{CC'}}{\overline{AC}} = \frac{\overline{MM'}}{\overline{AM}} = \frac{\gamma}{2}.$$

Correspondingly we find for the other diagonal $\overline{BD}$:

$$\epsilon_2 = -\frac{\gamma}{2}$$

and therefore

$$\epsilon_1 - \epsilon_2 = \gamma.$$

Turning the axes of coordinates in Fig. 184 by 45 deg, we get

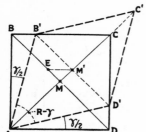

$$\sigma_x - \sigma_y = 2\tau_{xy} = 2\mu\frac{\partial\gamma}{\partial t} = 2\mu\frac{\partial}{\partial t}(\epsilon_1 - \epsilon_2).$$

In this system of coordinates we have:

$$\frac{\partial\epsilon_1}{\partial t} = \frac{\partial u}{\partial x},$$

and

$$\frac{\partial\epsilon_2}{\partial t} = \frac{\partial v}{\partial y},$$

Fig. 185.— Change in length of diagonals due to shear stresses.

so that

$$\sigma_x - \sigma_y = 2\mu\left(\frac{\partial u}{\partial x} - \frac{\partial v}{\partial y}\right). \qquad (3a)$$

In the same way we obtain

$$\sigma_x - \sigma_z = 2\mu\left(\frac{\partial u}{\partial x} - \frac{\partial w}{\partial z}\right). \qquad (3b)$$

The pressure in the fluid is now defined as being the negative of the mean value of the normal stress on a sphere of unit radius. The calculation of this value gives for the pressure[1]

$$p = -\tfrac{1}{3}(\sigma_x + \sigma_y + \sigma_z). \qquad (4)$$

Adding Eqs. (3a) and (3b), we obtain after some transformation

$$\sigma_x = -p - \frac{2}{3}\mu\left(\frac{\partial u}{\partial x} + \frac{\partial v}{\partial y} + \frac{\partial w}{\partial z}\right) + 2\mu\frac{\partial u}{\partial x}.$$

Similarly the following two equations are found:

$$\sigma_y = -p - \frac{2}{3}\mu\left(\frac{\partial u}{\partial x} + \frac{\partial v}{\partial y} + \frac{\partial w}{\partial z}\right) + 2\mu\frac{\partial v}{\partial y},$$

$$\sigma_z = -p - \frac{2}{3}\mu\left(\frac{\partial u}{\partial x} + \frac{\partial v}{\partial y} + \frac{\partial w}{\partial z}\right) + 2\mu\frac{\partial w}{\partial z}.$$

[1] The mean value of the normal stress on a sphere of unit radius can be found by first calculating the normal components of the stress on each

107. Relation between the Stress Tensor and the Velocity Tensor.—In the previous articles the relation between the state of stress and the rate of change of deformation for various simple cases has been discussed. It remains now to find the general expression for this relation containing as special cases all those previously considered.

The state of stress has been expressed by the general expression II. In an analogous manner an expression containing nine elements can be found for the velocity field. This expression will be built up of partial derivatives of the three components of velocity (see Art. 40). Since we have seen that the stress tensor is symmetrical, we have to find a relation between it and the symmetrical part of the velocity tensor. In Art. 44 it was found that the symmetrical part of the deformation tensor corresponds to the rate of change of elongation, while the antisymmetrical part is the expression for rotation. It is known that the sum of a tensor and its conjugate tensor, found by interchanging the rows and columns of the original tensor, represents a symmetrical tensor. Thus we have symbolically

---

surface element dA. This can be done by a scalar multiplication of the force $\mathbf{p}dA$ (in general oblique to the surface), with the radius $\mathbf{r}$ where $|\mathbf{r}| = 1$. Since the surface of a unit sphere is equal to 4π, the mean value becomes

$$\frac{1}{4\pi}\oiint \mathbf{r} \circ \mathbf{p}dA = \frac{1}{4\pi}\oiint \mathbf{r} \circ (dA_x\mathbf{p}_x + dA_y\mathbf{p}_y + dA_z\mathbf{p}_z);$$

or with the relations

$$dA = idA_x + jdA_y + kdA_z$$

and

$$\mathrm{II} = i\mathbf{p}_x + j\mathbf{p}_y + k\mathbf{p}_z$$

this becomes

$$\frac{1}{4\pi}\oiint \mathbf{r} \circ (dA \circ \mathrm{II}) = \frac{1}{4\pi}\oiint dA \circ (\mathbf{r} \circ \mathrm{II}),$$

since $\mathbf{r} \| dA$.
Applying Gauss's theorem this becomes

$$\frac{1}{4\pi}\oiint dA \circ (\mathbf{r} \circ \mathrm{II}) = \frac{1}{4\pi}\iiint \mathrm{div}\ (\mathbf{r} \circ \mathrm{II})dV = \frac{1}{3}\,\mathrm{div}\ (\mathbf{r} \circ \mathrm{II}),$$

considering that the volume of the unit sphere is equal to $\frac{4}{3}\pi$. Since any stress field can be considered homogeneous in a sufficiently small region, this is

$$\frac{1}{4\pi}\oiint \mathbf{r} \circ \mathbf{p}dA = \frac{1}{3}\,\mathrm{div}\ \mathbf{r} \circ \mathrm{II} = \frac{1}{3}(i \circ \mathbf{p}_x + j \circ \mathbf{p}_y + k \circ \mathbf{p}_z) = \frac{1}{3}(\sigma_x + \sigma_y + \sigma_z).$$

$$\begin{Bmatrix} \sigma_x & \tau_{xy} & \tau_{xz} \\ \tau_{xy} & \sigma_y & \tau_{yz} \\ \tau_{xz} & \tau_{yz} & \sigma_z \end{Bmatrix} = \mu \begin{Bmatrix} \dfrac{\partial u}{\partial x} & \dfrac{\partial v}{\partial x} & \dfrac{\partial w}{\partial x} \\[2mm] \dfrac{\partial u}{\partial y} & \dfrac{\partial v}{\partial y} & \dfrac{\partial w}{\partial y} \\[2mm] \dfrac{\partial u}{\partial z} & \dfrac{\partial v}{\partial z} & \dfrac{\partial w}{\partial z} \end{Bmatrix} + \mu \begin{Bmatrix} \dfrac{\partial u}{\partial x} & \dfrac{\partial u}{\partial y} & \dfrac{\partial u}{\partial z} \\[2mm] \dfrac{\partial v}{\partial x} & \dfrac{\partial v}{\partial y} & \dfrac{\partial v}{\partial z} \\[2mm] \dfrac{\partial w}{\partial x} & \dfrac{\partial w}{\partial y} & \dfrac{\partial w}{\partial z} \end{Bmatrix}$$

$$- \begin{Bmatrix} p & 0 & 0 \\ 0 & p & 0 \\ 0 & 0 & p \end{Bmatrix} - \frac{2}{3}\mu \begin{Bmatrix} \operatorname{div} \mathbf{w} & 0 & 0 \\ 0 & \operatorname{div} \mathbf{w} & 0 \\ 0 & 0 & \operatorname{div} \mathbf{w} \end{Bmatrix}$$

In the notations of vector analysis, this expression becomes

$$\Pi = \mu(\nabla \mathbf{w} + \mathbf{w}\nabla) - p - \tfrac{2}{3}\mu \operatorname{div} \mathbf{w}.$$

In order to recognize that the relations between the stress and the velocity given by this formula contain the special cases discussed before, it is only necessary to write the connection between the expressions having similar locations in the symbolic expression, for instance:

$$\sigma_x = 2\mu\frac{\partial u}{\partial x} - p - \frac{2}{3}\mu \operatorname{div} \mathbf{w}, \qquad \tau_{xy} = \mu\left(\frac{\partial v}{\partial x} + \frac{\partial u}{\partial y}\right), \text{ etc.}$$

108. The Equation of Navier-Stokes.—Now we have come to the point where the total surface force **G** can be expressed as a function of the rates of change of deformation. Taking first the x-components of **G**, we have

$$G_x = \left(\frac{\partial \sigma_x}{\partial x} + \frac{\partial \tau_{xy}}{\partial y} + \frac{\partial \tau_{xz}}{\partial z}\right)$$

and substituting into it the values for σ_x, τ_{xy}, τ_{xz}, found above, this becomes

$$G_x = -\frac{\partial p}{\partial x} + \mu\left(\frac{\partial^2 u}{\partial x^2} + \frac{\partial^2 u}{\partial y^2} + \frac{\partial^2 u}{\partial z^2}\right) + \frac{1}{3}\mu\frac{\partial}{\partial x}\left(\frac{\partial u}{\partial x} + \frac{\partial v}{\partial y} + \frac{\partial w}{\partial z}\right).$$

In a similar manner, the y- and z-components of **G** are found. Written in vector notation,

$$\mathbf{G} = \nabla \circ \Pi = \mu \nabla \circ (\nabla \mathbf{w} + \mathbf{w}\nabla) - \operatorname{grad} p - \tfrac{2}{3} \operatorname{grad} \operatorname{div} \mathbf{w}$$

and, since

$$\mu \nabla \circ (\nabla \mathbf{w} + \mathbf{w}\nabla) = \mu \nabla \circ \nabla \mathbf{w} + \mu \operatorname{grad} \operatorname{div} \mathbf{w},$$

this is

$$\mathbf{G} = -\operatorname{grad} p + \tfrac{1}{3}\mu \operatorname{grad} \operatorname{div} \mathbf{w} + \mu \Delta \mathbf{w}.$$

Substituting this expression into the fundamental equation derived at the beginning of this chapter, we find

$$\frac{D\mathbf{w}}{dt} = \mathbf{g} - \frac{1}{\rho} \operatorname{grad} p + \frac{1}{3}\nu \operatorname{grad} \operatorname{div} \mathbf{w} + \nu\Delta\mathbf{w}, \tag{5}$$

where $\nu = \mu/\rho$.

The x-component of this equation reads

$$\frac{\partial u}{\partial t} + u\frac{\partial u}{\partial x} + v\frac{\partial u}{\partial y} + w\frac{\partial u}{\partial z}$$

$$= X - \frac{1}{\rho}\frac{\partial p}{\partial x} + \frac{1}{3}\nu\frac{\partial}{\partial x}\left(\frac{\partial u}{\partial x} + \frac{\partial v}{\partial y} + \frac{\partial w}{\partial z}\right) + \nu\left(\frac{\partial^2 u}{\partial x^2} + \frac{\partial^2 u}{\partial y^2} + \frac{\partial^2 u}{\partial z^2}\right).$$

This differential equation, known as the equation of Navier-Stokes, is the fundamental equation of hydrodynamics. It is valid for compressible as well as for incompressible fluids. Since in the latter case div $\mathbf{w} = 0$, it is simplified to

$$\frac{D\mathbf{w}}{dt} = \mathbf{g} - \frac{1}{\rho} \operatorname{grad} p + \nu\Delta\mathbf{w}. \tag{6}$$

The difference between this equation and that of Euler for the frictionless fluid lies in the last term $\nu\Delta\mathbf{w}$, expressing the effect of friction.

The equation was first found by Navier[1] (1827) and Poisson[2] (1831). Their derivation was based on certain theories of the action of intermolecular forces. Without using hypotheses of this kind, St. Venant[3] (1843) and Stokes[4] (1845) found the same equation on the assumptions that the normal and shear stresses are linear functions of the deformation velocities (Newton's law of viscosity) and that the mean normal pressure is independent of the velocity of dilatation (for compressible fluids). The assumption, therefore, is that internal friction makes itself felt only when layers of the fluid slide relatively to each other but not, as in the case of a pure dilatation, when the volume of the fluid changes without sliding.

[1] NAVIER, Memoir on the Laws of Fluid Motion (French), *Mém. acad. sci.*, vol. 6, p. 389, 1827.

[2] POISSON, Memoir on the General Equations of the Equilibrium and the Motion of Solid Elastic Bodies and of Fluids (French), *J. école polytech.*, vol. 13, p. 1, 1831.

[3] ST. VENANT, *Compt. rend.* (French), vol. 17, p. 1240, 1843.

[4] STOKES, On the Theories of the Internal Friction of Fluids in Motion, *Trans. Cambridge Phil. Soc.*, vol. 8, 1845.

The question whether the hypotheses on which the above equation was derived are true can be decided only by experiment. The conclusions which have been drawn so far from the equation have been found to be in good agreement with the experimental facts. Especially the laws of laminar flow in straight tubes of circular cross section are a striking proof for the validity of the equation of Navier-Stokes for incompressible flow.[1]

However, it has to be considered that on account of the great mathematical difficulties not a single exact solution of the equation is known where the "convective" terms in their most general relation appear alongside the friction terms. With the exact solution of laminar flow in tubes, the convective terms vanish and the other known solutions where the convective terms are not zero[2] are always special cases where the velocities depend in a very simple manner on the coordinates.

109. Discussion of the Navier-Stokes Equation.—It is noted that the various cases of potential flow of an incompressible fluid are exact solutions of Navier-Stokes equation, since in those cases the friction term disappears. When Φ is the potential function, the equation for potential flow is

$$\Delta \Phi = 0,$$

so that

$$\text{grad } \Delta \Phi = \Delta \text{ grad } \Phi = 0,$$

or, since

$$\text{grad } \Phi = \mathbf{w},$$

we have

$$\Delta \mathbf{w} = 0.$$

Therefore in a potential flow, the shear stresses acting on any elementary fluid particle are in equilibrium in themselves. However, it is not possible to satisfy at the same time the two necessary boundary conditions, stating that the normal as well as the tangential velocity at a wall has to be zero. In case the normal component is given, the equation $\Delta \Phi = 0$ determines the tangential component uniquely. With a differ-

---

[1] For compressible fluids Navier's equation assumes that the rate of change of dilatation div $\mathbf{w}$ has no effect on the friction or on the pressures. Up to date no experimental proof or disproof for this assumption has been given.

[2] VON KÁRMÁN, TH., On Laminar and Turbulent Friction (German), *Z. angew. Math. Mech.*, vol. 1, p. 233, 1921.

ential equation of the second order in Φ, it is not possible to prescribe the value of this tangential component. In order to do that, a differential equation of higher order is necessary.

Such an equation can be obtained by introducing the stream function and eliminating the pressure. On page 116 it was shown that in the case of a homogeneous incompressible fluid the static pressure and the gravity force can be neglected, since the action of gravity on the individual elements of the fluid is eliminated by their buoyancy. This consideration naturally is restricted to cases where no free surfaces of the fluid occur. We therefore write

$$\frac{D\mathbf{w}}{dt} = -\frac{1}{\rho} \operatorname{grad} p + \nu\Delta\mathbf{w}. \tag{7}$$

In the case of two-dimensional motion, it is useful to introduce the stream function Ψ,

$$u = \frac{\partial\Psi}{\partial y}, \qquad v = -\frac{\partial\Psi}{\partial x},$$

by which the condition of continuity is automatically satisfied. Differentiating the x-component of Eq. (7) with respect to y and its y-component with respect to x and then subtracting these two from each other, the pressure is eliminated. After some transformation, we obtain

$$\frac{\partial}{\partial t}\Delta\Psi + u\frac{\partial}{\partial x}\Delta\Psi + v\frac{\partial}{\partial y}\Delta\Psi = \nu\Delta\Delta\Psi, \tag{8}$$

in which the meaning of $\Delta\Psi$ is $-\operatorname{rot}\mathbf{w}$. The left side of the equation is the total differential coefficient of $\Delta\Psi$. The equation expresses therefore the change in rotation of an element due to the friction. On account of the term on the right side, this equation is of the fourth order.

110. The Differential Equation of Creeping Motion.—On account of the great mathematical difficulties involved in integrating Eq. (7) or (8), it becomes necessary to introduce simplifying assumptions in order to obtain any solution at all. The difficulties are principally due to two facts: first, that the equation is quadratic so that no superposition of particular solutions is possible, and, second, that the terms proportional to ν are of a higher order. It has been seen before that if these friction terms are completely neglected, potential solutions are obtained which satisfy the complete differential equation but cannot be made to fit the general boundary conditions.

Limiting ourselves to cases where the viscosity is very large or where the velocity and the dimension and consequently the Reynolds' number, is very small, it is found that the quadratic term,

$$\mathbf{w} \circ \nabla \mathbf{w}$$

or

$$u\frac{\partial u}{\partial x}, \qquad v\frac{\partial u}{\partial y}, \qquad \text{etc.,}$$

becomes very small with respect to the friction term $\nu\Delta\mathbf{w}$. Stokes[1] has shown that Eq. (7) can be integrated in certain cases when the quadratic term is neglected. However, the approximate solution thus obtained has been found to agree with experimental facts only for very small values of the Reynolds' number. "Creeping" motions of this sort are encountered, for instance, in the rise of air bubbles in syrup or in the falling of the water particles constituting a cloud (not, however, in falling rain drops, because there the Reynolds' number is far too large).

In case of steady motion, Eq. (7) becomes

$$\mu\Delta\mathbf{w} = \text{grad } p,$$

i.e., $\Delta\mathbf{w}$ must be the gradient of a function of the coordinates only. The equation can be transformed somewhat by forming its rotation and thus eliminating the pressure. Considering that rot grad $p \equiv 0$, we obtain

$$\text{rot } \Delta\mathbf{w} = 0.$$

By writing

$$\mathbf{w} = \text{rot } \mathbf{A},$$

where $\mathbf{A}$ is an arbitrary vector, we automatically satisfy the continuity equation div $\mathbf{w} = 0$ (see Art. 47). Thus

$$0 = \text{rot } \Delta(\text{rot } \mathbf{A}) = \Delta(\text{rot rot } \mathbf{A}).$$

With the assumption div $\mathbf{A} = 0$, which is always permissible, we have further,

$$\text{rot rot } \mathbf{A} = -\Delta\mathbf{A},$$

so that the differential equation of creeping motion becomes

$$\Delta\Delta\mathbf{A} = 0.$$

After solutions of this differential equation have been found which also satisfy the boundary conditions, the pressure can be determined from the simplified equation of Navier-Stokes.

[1] STOKES, *Cambridge Phil. Trans.*, vol. 8, 1845; vol. 9, 1851; or *Papers*, vol. 1, p. 75.

Stokes has calculated the case of creeping motion about a sphere, and by integrating the pressure and the shear stresses on the entire surface he determined the resistance.

A more general case where the differential equation (7) remains linear occurs when the "local" acceleration $\partial \mathbf{w}/\partial t$ is the only term retained of the total differential coefficient. This is generally valid for cases of small oscillations where the quadratic term (convective acceleration) is small with respect

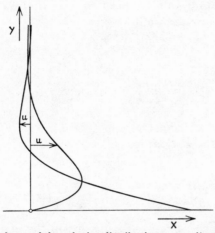

FIG. 186.—Two phases of the velocity distribution perpendicular to an oscillating wall in a viscous fluid at rest.

to the local acceleration. When, for instance, the displacement from the equilibrium position is given by

$$\xi = a \sin \alpha t,$$

we have

$$u = \frac{\partial \xi}{\partial t} = a\alpha \cos \alpha t$$

and

$$\frac{\partial u}{\partial t} = \frac{\partial^2 \xi}{\partial t^2} = -a\alpha^2 \sin \alpha t.$$

If l is a characteristic length of the phenomenon, $\partial u/\partial x$ is of the order u/l. Now $u(\partial u/\partial x)$ and u^2/l have the same dimension as $\partial u/\partial t$; however, on account of

$$\frac{u^2}{l} = \frac{a^2}{l}\alpha^2 \cos^2 \alpha t,$$

these two expressions can be neglected with respect to the local acceleration $\partial u/\partial t$, if a/l is small with respect to unity.

As an example we shall discuss the flow of a fluid in the neighborhood of a wall oscillating in the direction of its plane (Fig. 186). Let the displacement from the equilibrium position be

$$\xi = f(y)e^{i\alpha t}.$$

In this case, $u(\partial u/\partial x) = 0$ exactly, so that the calculations hold for large values of the amplitude a.

Since p is constant on account of the fact that no acceleration in the y-direction exists, we have

$$\frac{\partial u}{\partial t} = \nu\frac{\partial^2 u}{\partial y^2}.$$

Since

$$u = \frac{\partial \xi}{\partial t} = i\alpha f(y)e^{i\alpha t},$$

it follows that

$$i\alpha f(y) = \nu f''(y).$$

Assuming that $f(y) = Ce^{\beta y}$, we get

$$i\alpha = \nu\beta^2$$

and consequently

$$\beta = \pm\sqrt{\frac{i\alpha}{\nu}} = \pm(1 + i)\sqrt{\frac{\alpha}{2\nu}}.$$

Since $f(y) = 0$ for $y \to \infty$, we finally obtain the real expression

$$f(y) = e^{-y\sqrt{\frac{\alpha}{2\nu}}}\Big(A \cos y\sqrt{\frac{\alpha}{2\nu}} + B \sin y\sqrt{\frac{\alpha}{2\nu}}\Big),$$

where A and B are constants.

111. Oseen's Improvement of the Theory.—The state of flow in the vicinity of a sphere, and consequently its resistance, are given very accurately by Stokes's equation. However, the flow at larger distances away from it is found to be quite different from the calculated one. It can be shown that the rate of decrease with the distance from the sphere of the (very small) inertia forces is smaller than that of the viscosity forces. Although the viscosity forces are of exclusive importance in the vicinity of the sphere, this is not the case at sufficiently large distances from it. There the inertia forces cannot be neglected with respect to the viscosity forces. A fundamental improvement in the theory in this direction was made by Oseen.[1]

[1] Oseen, C. W., On the Formula of Stokes (German), *Arkiv. Math. Astron. Fysik*, vol. 6, 1910; vol. 7, 1911.

Lamb[1] has given a much simpler derivation of this work and also a physical interpretation of it. The influence of inertia is considered in so far that the quadratic term $\mathbf{w} \circ \nabla \mathbf{w}$ is replaced by $\mathbf{W} \circ \nabla \mathbf{w}$, where $\mathbf{W}$ is the velocity of the fluid at infinity with respect to the body. This entails the simplification of having the velocity $\mathbf{w}$ appear in a linear form. Since the inertia becomes of importance only at great distances from the sphere and since there $\mathbf{W}$ does not differ much from $\mathbf{w}$, Oseen's analysis considers the main effects of the "convective" acceleration. Stokes's calculation results in a flow picture symmetrically arranged in front of and behind the sphere, while Oseen's analysis leads to an unsymmetrical streamline picture. A definite difference between the streamlines of Stokes and those of Oseen can be detected only at some distance away from the sphere, whereas in its immediate neighborhood both solutions coincide.

The problem of two-dimensional creeping flow is determined by the differential equation

$$\Delta\Delta\Psi = 0,$$

where Ψ is the stream function. Since for the case of a circular *cylinder* this equation leads to a trivial result, Stokes did not succeed in finding a solution to that problem. Lamb found the solution later, considering the inertia term $\mathbf{W} \circ \nabla \mathbf{w}$ in practically the same manner as Oseen had done before in the case of the sphere.

---

[1] LAMB, H., On the Uniform Motion of a Sphere through a Viscous Fluid, *Phil. Mag.*, vol. 21 p. 120, 1911.

INDEX

A CATALOGUE OF SELECTED
DOVER SCIENCE BOOKS

A CATALOGUE OF SELECTED
DOVER SCIENCE BOOKS

WHAT IS SCIENCE? Norman R. Campbell. Scientific method, role of math, scientific laws, measurement, application of science, and similar topics. Used in hundreds of high schools, colleges. 186pp. 60043-2 Pa. $2.50

FADS AND FALLACIES IN THE NAME OF SCIENCE, Martin Gardner. Fair, witty appraisal of cranks and quacks of science: Atlantis, Lemuria, flat earth, Velikovsky, orgone energy, Bridey Murphy, medical fads, etc. 373pp. 20394-8 Pa. $3.00

THE RESTLESS UNIVERSE, Max Born. Nobel Laureate explains waves and particles, atomic structure, nuclear physics, similar topics. 132 illustrations. 7 "flip" sequences. 315pp. 6⅛ x 9¼. 20412-X Pa. $3.50

EINSTEIN'S THEORY OF RELATIVITY, Max Born. Finest semi-technical account; covers Einstein, Lorentz, Minkowski, and others, with much detail, much explanation of ideas and math not readily available elsewhere on this level. For student, nonspecialist. 376pp. 60769-0 Pa. $3.50

EINSTEIN: THE MAN AND HIS ACHIEVEMENT, edited by G.J. Whitrow. Insightful portrait covering all phases of Einstein's life and work. Excellent commentaries by Whitrow, W.B. Bonner, and D.W. Sciama; reminiscences by H.A. Einstein, Bertrand Russell, and 23 others. 94pp. USO 22934-3 Pa. $1.50

THE EVOLUTION OF SCIENTIFIC THOUGHT FROM NEWTON TO EINSTEIN, A. d'Abro. Evolution of classical physics into special and general relativity. Riemann, Weyl, Planck, Maxwell, Einstein. Stresses development for reader with some mathematical, physical background. 482pp. 20002-7 Pa. $5.00

THE RISE OF THE NEW PHYSICS, A. d'Abro. Most thorough exposition on semi-technical level of mathematical physics, Newton to Heisenberg, Dirac. Philosophies, causality, relativity, mechanics, electromagnetism, with higher mathematics developed within exposition for reader. 97 illustrations. 982pp.
20003-5, 20004-3 Pa., Two vol. set $10.00

GREAT IDEAS OF MODERN MATHEMATICS, Jagjit Singh. Winner of Unesco Kalinga Award explains verbally differential equations, matrices, groups, sets, mathematical logic, etc. and use in physics, astronomy, pure math. Excellent refresher, survey for layman, scientist in other area. 312pp. USCO 20587-8 Pa. $3.00

GREAT IDEAS OF OPERATIONS RESEARCH, Jagjit Singh. Non-technical, easily followed explanation of statistics, linear programming, game theory, queuing theory, Monte Carlo simulation, etc., and use. Includes case studies. 228pp.
EBE 21886-4 Pa. $3.50

GREAT IDEAS IN INFORMATION THEORY, LANGUAGE AND CYBERNETICS, Jagjit Singh. Clear, thorough, non-technical. Information flow, coding, analogue and digital computers, neural networks; work of von Neumann, Turing, McCulloch, Wiener. Some math for more advanced reader. 338pp.　　　EBE 21694-2 Pa. $3.50

GREAT IDEAS AND THEORIES OF MODERN COSMOLOGY, Jagjit Singh. Einstein's space-time, Milne's world-system, theories of Alfven, Eddington, Hoyle's "continuous creation" explained for layman. 276pp.　　　EBE 20925-3 Pa. $3.50

DIGITAL COMPUTER BASICS, U.S. Bureau of Naval Personnel. Fundamental concepts of electronic computers and data processing: operation, programming, analog-digital conversions, diagnostic maintenance routines. 161 figures. 231pp. 6½ x 9¼.　　　22480-5 Pa. $2.00

BASIC DATA PROCESSING, U.S. Bureau of Naval Personnel. Punch card language, working of computer, card sorters, input and output, data flow, programming, etc. Emphasis on actual commercial practice; 1970 rev. ed. 270 illustrations. 316pp. 6½ x 9¼.　　　20229-1 Pa. $3.00

INTRODUCTION TO ELECTRONICS, U.S. Bureau of Naval Personnel. Basic concepts, techniques, equipment on elementary level. Power supplies, electron tubes in circuits, operation of transistors, radio, radar, etc. 145pp. 6½ x 9¼.
21283-1 Pa. $1.50

BASIC ELECTRONICS, U.S. Bureau of Naval Personnel. Using nothing more advanced than elementary math and electricity, this manual covers electron tubes, circuits, antennas, AM, FM, and CW transmission and receiving, etc. 560 illustrations. 567pp. 6½ x 9¼.　　　21076-6 Pa. $5.00

BASIC ELECTRICITY, U.S. Bureau of Naval Personnel. Originally a training course, best non-technical coverage. Batteries, circuits, conductors, AC and DC, inductance and capacitance, generators, motors, transformers, amplifiers, etc. Many questions, answers. 349 illustrations. 448pp. 6½ x 9¼.　　　20973-3 Pa. $4.50

BASIC THEORY AND APPLICATION OF TRANSISTORS, U.S. Department of the Army. Fundamental theory, applications for persons with minimal background. Amplifiers, analysis and comparison, modulation, semiconductors, etc. 263pp. 6½ x 9¼.　　　20380-8 Pa. $3.00

THE DIVINE PROPORTION: A STUDY IN MATHEMATICAL BEAUTY, H.E. Huntley. "Divine proportion" or "golden ratio" in poetry, Pascal's triangle, philosophy, psychology, music, mathematical figures, etc. Excellent bridge between science and art. 58 figures. 185pp.　　　22254-3 Pa. $2.50

A SHORT ACCOUNT OF THE HISTORY OF MATHEMATICS, W.W. Rouse Ball. One of clearest, most authoritative surveys from the Egyptians and Phoenicians through 19th-century figures such as Grassman, Galois, Riemann. Fourth edition. 522pp.
20630-0 Pa. $5.00

A CONCISE HISTORY OF MATHEMATICS, Dirk J. Struik. The best brief history of mathematics. Stresses origins and covers every major figure from ancient Near East to 19th century. Third revised edition. 41 illustrations. 195pp.

EUK 60255-9 Pa. $2.50

THE HISTORY OF THE CALCULUS AND ITS CONCEPTUAL DEVELOPMENT, Carl B. Boyer. Origins in antiquity, medieval contributions, work of Newton, Leibniz, rigorous formulation. Treatment is verbal. Foreword by R. Courant. 346pp.

60509-4 Pa. $4.00

THE THIRTEEN BOOKS OF EUCLID'S ELEMENTS, translated with introduction and commentary by Sir Thomas L. Heath. Definitive edition. Textual and linguistic notes, mathematical analysis, 2500 years of critical commentary. Do not confuse with abridged school editions. Total of 1,414pp.

60088-2, 60089-0, 60090-4 Pa., Three vol. set $15.00

NON-EUCLIDEAN GEOMETRY, Roberto Bonola. Standard historical and critical survey; many systems not usually represented. Easily followed by nonspecialist. Also includes Bolyai's "The Science of Absolute Space" and Lobachevski's "The Theory of Parallels." 181 diagrams. Total of 431pp.   60027-0 Pa. $4.50

THE GEOMETRY OF RENÉ DESCARTES, René Descartes. The great work founded analytical geometry. Original French text, Descartes' own diagrams, together with definitive Smith-Latham translation. 244pp.   60068-8 Pa. $3.00

DIALOGUES CONCERNING TWO NEW SCIENCES, Galileo Galilei. Encompassing 30 years of experiment and thought, these dialogues deal with geometric demonstrations of fracture of solid bodies, cohesion, leverage, speed of light and sound, pendulums, falling bodies, accelerated motion, etc. 300pp.   60099-8 Pa. $3.50

OPTICKS, Sir Isaac Newton. Survey of 18th-century knowledge, Newton's own experiments with spectroscopy, colors, lenses, reflection, refraction, etc. in language the layman can follow. Foreword by Albert Einstein. Total of 532pp.

60205-2 Pa. $4.00

A TREATISE ON ELECTRICITY AND MAGNETISM, James Clerk Maxwell. Important foundation work of modern physics. Brings to final form Maxwell's theory of electromagnetism and rigorously derives his general equations of field theory. Total of 1,084pp.   EUK 60636-8, 60637-6 Pa., Two vol. set $10.00

THE HISTORICAL BACKGROUND OF CHEMISTRY, Henry M. Leicester. Unique approach developed in terms of evolution of ideas, not individual biography, and related to events of world history. Concentrates on formulation of a coherent set of chemical laws. 15 figures. 260pp.   61053-5 Pa. $3.00

A SHORT HISTORY OF ASTRONOMY, Arthur Berry. Earliest times through the 19th century. Individual chapters on Copernicus, Tycho Brahe, Galileo, Kepler, Newton, etc. Non-technical, but precise. For specialist or layman. 104 illustrations, 9 portraits. 440pp.   20210-0 Pa. $5.00

COMPUTATIONAL METHODS OF LINEAR ALGEBRA, V.N. Faddeeva. Classical and modern Russian computational methods including work of A.N. Krylov, A.M. Danilevsky, D.K. Faddeev and others. New bibliography. 252pp.
60424-1 Pa. $3.50

LINEAR ALGEBRA AND GROUP THEORY, V.I. Smirnov. "Concrete" approach for undergraduates, working physicists and engineers. Determinants and systems of equations; matrix theory; group theory. Problems, answers. Adapted and edited by Richard Silverman. 464pp.
62624-5 Pa. $4.50

INTRODUCTION TO THE THEORY OF GROUPS OF FINITE ORDER, Robert D. Carmichael. Progresses in easy steps from sets, groups, permutations, isomorphism, through important types of groups. 783 problems and exercises. 447pp. 60300-8 Pa. $4.50

THE THEORY OF GROUPS AND QUANTUM MECHANICS, Hermann Weyl. Landmark of mathematics, devoted to consistent and systematic application of group theory to quantum mechanics. Classical theory; relations between mathematical and physical theories. Total of 444pp.
60269-9 Pa. $4.50

AN INTRODUCTION TO THE THEORY OF LINEAR SPACES, Georgi E. Shilov. Complete introduction to linear algebra and functional analysis. Determinants, linear spaces, systems of linear equations, linear functions of a vector argument, orthogonalization and measurement of volume, more. Numerous problems. 9 figures. 310pp.
63070-6 Pa. $4.00

AN INTRODUCTION TO THE THEORY OF STATIONARY RANDOM FUNCTIONS, A.M. Yaglom, translated by R.A. Silverman. Classic study by foremost Russian authority covers general theory and Wiener-Kolmogorov theory of extrapolation and interpolation. 235pp.
60579-5 Pa. $3.50

THE THEORY OF FUNCTIONS, Konrad Knopp. "An excellent introduction ... remarkably readable, concise, clear, rigorous" —the Journal of the American Statistical Association. Widely used as a college text.

ELEMENTS OF THE THEORY OF FUNCTIONS, Konrad Knopp. General background: complex numbers, linear functions, sets and sequences, conformal mapping. Detailed proofs. 140pp.
60154-4 Pa. $2.00

THEORY OF FUNCTIONS, PART I, Konrad Knopp. Full, rigorous demonstrations of the general foundations: integral theorems, series, the expansion of analytic functions. 146pp.
60156-0 Pa. $2.00

THEORY OF FUNCTIONS, PART II, Konrad Knopp. Single- and multiple-valued functions; full presentation of the most characteristic and important types. Proofs fully worked out. 150pp.
60157-9 Pa. $2.00

PROBLEM BOOK IN THE THEORY OF FUNCTIONS, I, Konrad Knopp. More than 300 problems for use with "Theory of Functions, I." Detailed solutions. 126pp.
60158-7 Pa. $2.00

PROBLEM BOOK IN THE THEORY OF FUNCTIONS, II, Konrad Knopp. Over 230 problems for "Theory of Functions, II" or comparable text. Full solutions. 138pp.
60159-5 Pa. $2.00

SUMMATION OF SERIES, L.B.W. Jolley. 1,146 common series collected, summed and grouped for easy reference into 33 categories. References to further information. Only work of its kind now available. Revised, enlarged 2nd edition. 251pp.
60023-8 Pa. $3.00

CONFORMAL MAPPING, Zeev Nehari. A combined theoretical and practical approach that covers functions of a complex variable (one semester) and conformal mapping (one-year graduate course). Only prerequisite is a good working knowledge of advanced calculus. 396pp.
61137-X Pa. $5.00

INTRODUCTION TO BESSEL FUNCTIONS, Frank Bowman. Properties and applications: from Bessel functions of zero order to Bessel's solution to Kepler's problem, definitive integrals, asymptotic expansion, etc. Math above calculus developed within text. 636 problems. 135pp.
60462-4 Pa. $2.50

INTRODUCTORY REAL ANALYSIS, A.N. Kolmogorov, S.V. Fomin. Translated and edited by Richard A. Silverman. Self-contained, evenly paced introduction to real and functional analysis. Useful for self-study or as a basic one-year course. Some 350 problems. 403pp.
61226-0 Pa. $5.00

CALCULUS OF VARIATIONS, Robert Weinstock. Basic introduction covering isoperimetric problems, geometrical optics, Fermat's principle, dynamics of particles, Sturm-Liouville eigenvalue-eigenfunction problem, theory of elasticity, quantum mechanics, electrostatics. Exercises throughout. 326pp.
63069-2 Pa. $4.00

INTRODUCTION TO THE THEORY OF FOURIER'S SERIES AND INTEGRALS, Horatio S. Carslaw. Basic introduction to theory of infinite series and integrals, with special reference to Fourier's series. An important class text. 84 examples. 368pp.
60048-3 Pa. $4.00

INFINITE SEQUENCES AND SERIES, Konrad Knopp. Careful presentation of fundamentals of the theory by one of the finest modern expositors of higher mathematics. Translated by Frederick Bagemihl. 186pp.
60153-6 Pa. $2.50

INTRODUCTION TO NONLINEAR DIFFERENTIAL AND INTEGRAL EQUATIONS, Harold T. Davis. Definition of field through 1960, emphasizing work of Poincaré, Liapounoff, Painlevé, and Goursat. Mathematics held to moderate level of difficulty. 137 problems. 566pp.
60971-5 Pa. $5.50

ORDINARY DIFFERENTIAL EQUATIONS, Edward L. Ince. Explains and analyzes theory of ordinary differential equations in real and complex domains. "Highly recommended," Electronics Industries. 558pp.
60349-0 Pa. $6.00

DIFFERENTIAL ALGEBRA, Joseph F. Ritt. Modern coverage of concrete problems in theory of differential equations. Explains concepts of abstract algebra involved and reviews work of Raudenbush, Strodt, Kolchin, Levi, Gourin, Cohn. 184pp. $6\frac{1}{8}$ x $9\frac{1}{4}$. 61666-5 Pa. $2.50

NUMERICAL SOLUTION OF DIFFERENTIAL EQUATIONS, William E. Milne. First half demonstrates 13 methods of numerically solving ordinary equations; second half, partial equations. Practical examples throughout. Basis of computer operation with differential equations. 359pp. 62437-4 Pa. $4.00

THEORY OF SETS, E. Kamke. Lucid introduction to theory of sets, surveying discoveries of Cantor, Russell, Weierstrass, Zermelo, Bernstein, Dedekind, etc. Assumes college algebra. Translated by Frederick Bagemihl. 144pp. 60141-2 Pa. $2.00

ON INVARIANTS AND THE THEORY OF NUMBERS, Leonard E. Dickson. Dickson's theory of invariants for the modular forms and linear transformations employed in the theory of numbers. 110pp. 61667-3 Pa. $2.00

ELEMENTARY CONCEPTS OF TOPOLOGY, Paul Alexandroff. Elegant, intuitive approach to topology from set-theoretic topology to Betti groups; how concepts of topology are useful in math and physics. 25 figures. 57pp. 60747-X Pa. $1.50

PRINCIPLES OF NUMERICAL ANALYSIS, Alston S. Householder. General topics of solution of finite systems of linear and non-linear equations and approximate representation of functions. Reference for intermediate, advanced mathematicians, computer scientists. 54 problems. 274pp. 61116-7 Pa. $4.00

MATHEMATICAL TABLES AND FORMULAS, Robert D. Carmichael and Edwin R. Smith. All tables necessary for college algebra and trigonometry. Logarithms, sines, tangents, trig functions, powers, roots, reciprocals, exponential and hyperbolic functions, formulas and theorems. 269pp. 60111-0 Pa. $2.25

HANDBOOK OF MATHEMATICAL FUNCTIONS WITH FORMULAS, GRAPHS, AND MATHEMATICAL TABLES, edited by Milton Abramowitz and Irene A. Stegun. Vast compendium: 29 sets of tables, some to as high as 20 places. 1,046pp. 8 x $10\frac{1}{2}$. 61272-4 Pa. $9.95

TABLES OF FUNCTIONS: WITH FORMULAE AND CURVES, Eugene Jahnke and Fritz Emde. Revised fourth edition, with 76-page appendix. Sine, cosine; error integral; Riemann-Zeta function; exponential function; much more. 212 figures. 382pp. USO 60133-1 Pa. $4.50

AN ELEMENTARY INTRODUCTION TO THE THEORY OF PROBABILITY, B.V. Gnedenko and A. Ya. Khinchin. Excellent, highly accurate layman's introduction, extremely thorough within its range. Mathematics held to elementary level. 130pp. 60155-2 Pa. $2.75

SELECTED PAPERS ON NOISE AND STOCHASTIC PROCESSES, edited by Nelson Wax. Six papers by Chandrasekhar, Doob, Kac, Ming, Ornstein, Rice, and Uhlenbeck introduce advanced noise theory and fluctuation phenomena. 19 figures. 337pp. 6⅛ x 9¼. 60262-1 Pa. $4.00

MATHEMATICAL FOUNDATIONS OF INFORMATION THEORY, A. Ya. Khinchin. Comprehensive, rigorous introduction to work of Shannon, McMillan, Feinstein and Khinchin. Translated by R.A. Silverman and M.D. Friedman. 120pp.
60434-9 Pa. $2.00

ELEMENTARY STATISTICS: WITH APPLICATIONS IN MEDICINE AND THE BIOLOGICAL SCIENCES, Frederick E. Croxton. Fundamental techniques, methods of elementary statistics. Useful to readers in all fields. 376pp. 60506-X Pa. $3.00

STATISTICS MANUAL, Edwin L. Crow, et al. Comprehensive, practical collection of classical and modern methods prepared by U.S. Naval Ordnance Test Station. Stress on use. Basics of statistics assumed. 288pp. 60599-X Pa. $3.00

THE FOUNDATION OF STATISTICS, Leonard J. Savage. Classic analysis of foundation of statistics and development of personal probability, one of the greatest controversies in modern statistical thought. Revised edition with supplemental bibliography. 376pp. 62349-1 Pa. $3.50

VECTOR AND TENSOR ANALYSIS, George E. Hay. Clear introduction: simple definitions through oriented Cartesian vectors, Christoffel symbols, solenoidal tensors, applications. Many worked problems. 193pp. 60109-9 Pa. $3.00

CONTRIBUTIONS TO THE FOUNDING OF THE THEORY OF TRANSFINITE NUMBERS, Georg Cantor. The famous articles, 1895-97, that founded a new branch of mathematics. Translated with 82-page introduction by P. Jourdain. Still very good introduction. 211pp. 60045-9 Pa. $3.00

COLLECTED MATHEMATICAL PAPERS, George D. Birkhoff. Birkhoff (1888-1944) was one of world's greatest mathematicians, extremely prolific. These three volumes contain virtually all his published work in many areas of mathematics and mathematical physics, including the important relativity. A wealth of first-rate material. 2,700pp.
61955-9, 61956-7, 61957-5 Clothbd., Three vol. set $37.50

SCIENTIFIC PAPERS BY LORD RAYLEIGH. 446 papers cover the full working life of great British physicist, optics, gases, hydrodynamics, capillary action, thermodynamics, electricity, elastic solids, etc. Many extremely important contributions explained with model clarity. Total of 3,777pp. 6⅛ x 9¼.
61213-9, 61214-7, 61215-5 Clothbd., Three vol. set $45.00

INTRODUCTION TO CHEMICAL PHYSICS, John C. Slater. A work intended to bridge the gap between chemistry and physics. Level is advanced undergraduate to graduate; theoretical physics held to minimum. 522pp. 62562-1 Pa. $5.00

MATHEMATICAL PHYSICS, Donald H. Menzel. Mathematical techniques vital in classical mechanics, electromagnetic theory, quantum theory, relativity presented for junior, senior courses. 193 problems, answers. 412pp. 60056-4 Pa. $4.00

BASIC THEORIES OF PHYSICS: Mechanics and Electrodynamics, Peter G. Bergmann. Critical examination, especially useful in examining basic framework and methodology. Excellent supplement to courses, textbooks. Relatively advanced. 260pp. 60969-3 Pa. $3.00

FUNDAMENTAL FORMULAS OF PHYSICS, edited by Donald H. Menzel. Individual chapters, with full texts by leading authorities, cover statistics, physical constants, mechanics, relativity, aerodynamics, thermodynamics, electronics, acoustics, optics, quantum mechanics, etc. Special chapters on physical chemistry, astrophysics, meteorology and biophysics. Indispensable. Total of 848pp. 60595-7, 60596-5 Pa., Two vol. set $9.00

THE THEORY OF SOUND, J.W.S. Rayleigh. Valuable classic by great Nobel Laureate. Sums up previous research; Rayleigh's original contributions. New introduction and bibliography by R.B. Lindsay. Total of 1,042pp. 60292-3, 60293-1 Pa., Two vol. set $10.00

WAVE PROPAGATION IN PERIODIC STRUCTURES, Leon Brillouin. Modern classic in mathematical physics presents a general method for solving problems in areas from solid state physics to X-rays, rest rays, electrical engineering, etc. 255pp. 60034-3 Pa. $2.75

OPTICAL ROTATORY POWER, T. Martin Lowry. The classical work in light dispersion, from Biot to Born. Early apparatus, experiments and discoveries, applications. Basic background text for advanced study. 483pp. 61197-3 Pa. $5.00

AN INTRODUCTION TO THE LUMINESCENCE OF SOLIDS, Humboldt W. Leverenz. Introductory description of phosphors: preparation, composition, structure, physical characteristics, applications. Invaluable for illumination, detection of radiation. 569pp. 62028-X Pa. $6.00

ELECTROMAGNETISM, John C. Slater and Nathaniel H. Frank. Introductory study by leading men in the field supplies basic material on electrostatics, magnetostatics, electromagnetic theory. Calculus and differential equations required. Problems. 240pp. 62263-0 Pa. $3.00

HYDRODYNAMICS, Sir Horace Lamb. Standard reference and study work, almost inexhaustible in coverage of classical material. Unexcelled for fundamental theorems, equations, detailed methods of solution. 6th enlarged edition. 738pp. 6 x 9. USO 60256-7 Pa. $7.50

AN ELEMENTARY TREATISE ON THEORETICAL MECHANICS, Sir James Jeans. Great scientific expositor in remarkably clear presentation. Emphasizes physical principles rather than mathematics or applications. Hundreds of worked problems. 364pp. 61839-0 Pa. $2.75

A Treatise on the Mathematical Theory of Elasticity, A.E.H. Love. Most complete treatment of classical elasticity in a single volume. Coverage of stress, strain, bending, torsion, gravitational effects, etc. is rigorous and detailed. 643pp.
60174-9 Pa. $5.00

The Principle of Relativity, Albert Einstein et al. Eleven most important original papers on special and general theories. Seven by Einstein, two by Lorentz, one each by Minkowski and Weyl. All translated, unabridged. 216pp.
60081-5 Pa. $2.50

Principles of Quantum Mechanics, William V. Houston. Evidence for quantum theory, postulates of quantum mechanics, applications in spectroscopy, collision problems, electrons, similar topics. Uses Schroedinger's wave mechanics. 288pp.
60524-8 Pa. $3.50

Physical Principles of the Quantum Theory, Werner Heisenberg. Nobel Laureate discusses quantum theory, uncertainty, wave mechanics, work of Dirac, Schroedinger, Compton, Wilson, Einstein, etc. Middle, non-mathematical level. 184pp.
60113-7 Pa. $2.50

Selected Papers on Quantum Electrodynamics, edited by Julian Schwinger. 34 papers by Foley, Fermi, Heisenberg, Dyson, Weisskopf, Oppenheimer, Pauli, Schwinger, Klein and others. 3 papers in German, 1 each in French and Italian, balance in English. 424pp. $6\frac{1}{8}$ x $9\frac{1}{4}$.
60444-6 Pa. $4.50

Methods of Quantum Field Theory in Statistical Physics, A.A. Abrikosov et al. Translated and edited by Richard A. Silverman. Introduction to the many-body theory and its ramifications by three internationally known members of Academy of Sciences, U.S.S.R. 352pp.
63228-8 Pa. $5.00

Atomic Spectra and Atomic Structure, Gerhard Herzberg. One of best introductions to atomic spectra and their relationship to structure; especially for specialist in other fields. Treatment is physical rather than mathematical. 80 illustrations. 257pp.
60115-3 Pa. $3.00

Introduction to Chemical Physics, John C. Slater. Text in three parts: thermodynamics, statistical mechanics and kinetic theory; gases, liquids and solids; and atoms, molecules and the structure of matter. Theoretical physics held to minimum. 522pp.
62562-1 Pa. $5.00

The Nuclear Properties of the Heavy Elements, Earl K. Hyde, et al. Landmark investigation of nuclear instability and radioactivity for lead through element 103. Synthesizes, evaluates literature and data on Systematics of Nuclear Structure and Radioactivity (Vol. I), Detailed Radioactivity Properties (Vol. II), and Fission Phenomena (Vol. III). Total of 1,790pp.
62805-1, 62806-X, 62807-8 Clothbd., Three vol. set $45.00

SOLID STATE THEORY, Mendel Sachs. Nature of matter in solid state phase explained in general physical principles. Basically a graduate textbook, gives theoretical background to solid state physics, tangential fields. 350pp.

61772-6 Pa. $4.50

MATHEMATICAL FOUNDATIONS OF STATISTICAL MECHANICS, A. Ya. Khinchin. Phase space, ergodic problems, theory of probability, central limit theorem, ideal monatomic gas, dispersion and distribution of sum functions. Rigorous treatment and excellent analytical tools. 179pp.

60147-1 Pa. $3.00

INVESTIGATIONS ON THE THEORY OF THE BROWNIAN MOVEMENT, Albert Einstein. Five papers (1905-8) investigating dynamics of Brownian motion and evolving elementary theory. Notes by R. Fürth discuss history, analyze text and significance. 122pp.

60304-0 Pa. $2.25

THE DYNAMICAL THEORY OF GASES, Sir James Jeans. Great classic which introduced the notion of dissipation in kinetic theory. Treatment up through radiation and quantum theory. Organized so nonmathematical readers can follow development. 444pp. $6^{1}/_{8}$ x $9^{1}/_{4}$.

60136-6 Pa. $4.00

THERMODYNAMICS, Enrico Fermi. A classic of modern science. Clear, organized treatment of systems, first and second laws, entropy, thermodynamic potentials, gaseous reactions, dilute solutions, entropy constant. No math beyond calculus required. Problems. 160pp.

60361-X Pa. $2.50

CELESTIAL OBJECTS FOR COMMON TELESCOPES, T.W. Webb. The most used book in amateur astronomy: inestimable aid for locating and identifying nearly 4,000 celestial objects. Edited, updated by Margaret W. Mayall. 77 illustrations. Total of 645pp.

20917-2, 20918-0 Pa., Two vol. set $6.00

LUNAR ATLAS, Dinsmore Alter. Magnificent photographs of the lunar surface, both overall and closeups. Text gives lunar geography, explanation of surface features, photographic data. Preface by Harold Urey. 219 photographs. 341pp. $8^{3}/_{8}$ x $10^{1}/_{4}$.

21701-9 Pa. $6.00

INTRODUCTION TO ASTROPHYSICS: THE STARS, Jean Dufay. Best guide to observational astrophysics in English. Bridges the gap between elementary popularizations and advanced technical monographs. Translated by Owen Gingerich. 164pp.

60771-2 Pa. $2.50

RADIATIVE TRANSFER, Subrahmanyan Chandrasekhar. Foundation for analysis of stellar atmospheres, planetary illumination, sky radiation; physical interest for problems analogous to diffusion of neutrons. 393pp.

60590-6 Pa. $4.00

AN INTRODUCTION TO THE STUDY OF STELLAR STRUCTURE, S. Chandrasekhar. Rigorous examination, using both classical and modern mathematical methods, of relationship between loss of energy, mass and radius of stars in a steady state. 509pp.

60413-6 Pa. $4.50

MICROWAVE SPECTROSCOPY, C.H. Townes & A.L. Schawlow. Systematic, comprehensive account of theory, experimental data, experimental know-how developed in recent years. 18 chapters, can be read individually or as continuous text. 190 tables and figures. 698pp. 61798-X Pa. $7.50

THE PRINCIPLES OF ELECTROCHEMISTRY, Duncan A. MacInnes. Develops the basic equations for nearly all subfields from first principles referring at all times to the soundest theories and results in the literature. 478pp. 60052-1 Pa. $5.00

APPLIED BESSEL FUNCTIONS, Frederick E. Relton. Basic properties, plus applications to lengthening pendulum vibrations, curvature of railroad track, tidal motion, uniform chain, etc. Some differential equations required. 213 exercises. 191pp. 61511-1 Pa. $2.50

AN INTRODUCTION TO THE USE OF GENERALIZED COORDINATES IN MECHANICS AND PHYSICS, William E. Byerly. Clear, thorough exposition of coordinates in general systems given by Lagrange and Hamilton. Concrete illustrative examples. Problems. 118pp. 61362-3 Pa. $2.00

THEORY OF FLIGHT, Richard von Mises. Introduction to physical phenomena and mathematical concepts of fluid dynamics. Widely recommended for clarity, though limited to incompressible fluids. 629pp. 60541-8 Pa. $7.00

THEORY OF WING SECTIONS: INCLUDING A SUMMARY OF AIRFOIL DATA, Ira H. Abbott and A.E. von Doenhoff. Concise compilation of subsonic aerodynamic characteristics of modern NASA wing sections, plus description of theory. 350pp. of tables. 693pp. 60586-8 Pa. $6.00

FUNDAMENTALS OF HYDRO- AND AEROMECHANICS, Ludwig Prandtl and O.G. Tietjens. Tietjens' famous expansion of Prandtl's lectures: statics and kinematics of liquids and gases, dynamics of non-viscous liquids. Proofs use vector analysis. 270pp. 60374-1 Pa. $3.50

APPLIED HYDRO- AND AEROMECHANICS, Ludwig Prandtl and O.G. Tietjens. Methods valuable to engineers: flow in pipes, boundary layers, airfoil theory, entry conditions, turbulent flow, etc. Translated by J.P. Den Hartog. 311pp. 60375-X Pa. $4.50

WATERHAMMER ANALYSIS, John Parmakian. An important contribution to hydraulic engineering: an excellent exposition of the graphical method for solving waterhammer problems. 43 problems, solutions. 161pp. 61061-6 Pa. $3.00

NONSTEADY DUCT FLOW: WAVE DIAGRAM ANALYSIS, George Rudinger. Consistent set of computing procedures for the numerical analysis of nonsteady duct flow. Convenient manual or guide. 296pp. 62020-4 Pa. $3.50

TWO-DIMENSIONAL FIELDS IN ELECTRICAL ENGINEERING, L.V. Bewley. Covers field of flow, general theorems of mathematical physics, conformal mapping, method of images, freehand flux plotting, mechanical forces. Study problems. 204pp.
61118-3 Pa. $3.00

TRANSMISSION LINES, ANTENNAS AND WAVE GUIDES, Ronold W.P. King, et al. Non-mathematical, descriptive approach to transmission line, antennas, etc. for radio amateurs, practical engineers. 83 problems, many answered. 347pp.
61343-7 Pa. $3.00

EARTH CONDUCTION EFFECTS IN TRANSMISSION SYSTEMS, Erling D. Sunde. Basic electromagnetic concepts, testing earth resistivity, grounding arrangements, earth-return conductors, corrosion and lightning protection. 370pp.
61891-9 Pa. $4.00

POWER SYSTEM STABILITY: SYNCHRONOUS MACHINES, Edward W. Kimbark. Discusses such effects as saliency, damping, saturation and highspeed excitation as well as characteristics of machines themselves. Many worked examples. Problems. 322pp.
61885-4 Pa. $4.50

MECHANICS, J.P. Den Hartog. A classic introductory text or refresher. Hundreds of applications and design problems illuminate fundamentals of trusses, jacks, hoists, loaded beams and cables, gyroscopes, etc. 334 answered problems. 462pp.
60754-2 Pa. $4.50

STRENGTH OF MATERIALS, J.P. Den Hartog. Full, clear treatment of basic material (tension, torsion, bending, compound stress, deflection of beams), plus advanced material on engineering methods, applications. 350 answered problems. 323pp.
60755-0 Pa. $3.50

PHOTOELASTICITY: PRINCIPLES AND METHODS, H.T. Jessop and F.C. Harris. An introduction to general and modern developments in 2- and 3-dimensional stress analysis techniques. Advanced mathematical treatment in appendices. 184pp. 6⅛ x 9¼.
USO 60720-8 Pa. $3.00

OPTICS AND OPTICAL INSTRUMENTS: AN INTRODUCTION, B.K. Johnson. Telescopes, microscopes, photographic lenses and optical projection systems illustrate practical application of optics. How to set up working models, theoretical principles. Only elementary math required. 234 diagrams. 224pp.
60642-2 Pa. $3.50

THE PHYSICS OF MODERN ELECTRONICS, W.A. Günther. Physical theory behind the new types of oscillators: plasma cells, piezoelectric transducers, MHD generator, atomic clocks, solid state masers, the laser plus older devices. No mathematical training required. 337pp.
61749-1 Pa. $3.50

FREQUENCY ANALYSIS, MODULATION AND NOISE, Stanford Goldman. Well-written discussion of Fourier analysis and its applications to radio. Mathematical proofs developed only where necessary. 434pp.
61845-5 Pa. $4.00

HIGH-SPEED ANALOG COMPUTERS, Rajko Tomovic and Walter J. Karplus. Introduction to devices and circuits of computer facilities, and survey of applications to engineering problems. 255pp. 62564-8 Pa. $3.50

DIGITAL COMPUTER PROGRAMMING, Bing H. Lieu. Covers basic computer concepts, the essence of FORTRAN, computer hardware, number systems, numerical considerations, debug and check-out, and computer application in engineering analysis. 19 problems with answers. 38 figures. 228pp. 22948-3 Pa. $4.00

INTRODUCTION TO THE STATISTICAL DYNAMICS OF AUTOMATIC CONTROL SYSTEMS, V.V. Solodovnikov. Theory of linear analysis, statistics of random signals, theory of linear prediction and filtering. For advanced and graduate-level students. 307pp. 60420-9 Pa. $3.50

COMMUNICATION NETS, Leonard Kleinrock. Investigation of stochastic message flow and delay. Provides basis for understanding communication networks in realistic situations. 55 figures. 209pp. 61105-1 Pa. $3.00

FUNDAMENTAL THEORY OF SERVOMECHANISMS, LeRoy A. MacColl. Clear, elementary treatment of the mathematical theory of servomechanisms for students, designers in need of mathematical instruction. Based on frequency-domain approach. 130pp. 61916-8 Pa. $3.50

ELECTROMECHANICAL POWER CONVERSION, Enrico Levi, Marvin Panzer. New approach to transformation of mechanical to electrical power and vice versa. Natural phenomena involved, fission. About 1 semester course. 133 problems, solutions. 633pp. 60592-2 Pa. $8.00

OPERATIONAL METHODS IN NONLINEAR MECHANICS, Louis A. Pipes. A collection of problems involving nonlinear differential equations and methods for their solution especially pertinent to nonlinear electric circuit theory, servo-mechanisms. 99pp. 61441-7 Pa. $2.00

THE THEORY OF OPTIMUM NOISE IMMUNITY, Vladimir A. Kotel'nikov. Performance of communication systems in presence of additive Gaussian noise; basic work in information theory. Includes author's important contributions. 140pp. $7\frac{1}{8}$ x 10. 61952-4 Pa. $3.50

FUNDAMENTALS OF ASTRODYNAMICS, Roger Bate, et al. Most modern approach to teaching astronautics and aerospace engineering, developed by U.S. Air Force Academy. Designed as a first course. Problems; exercises. Numerous illustrations. 455pp. 60061-0 Pa. $5.00

GASEOUS CONDUCTORS, James D. Cobine. An indispensable reference for radio engineers, physicists and lighting engineers. Physical backgrounds, theory, applications in rectifiers, oscillographs, etc. 83 problems. 606pp. 60442-X Pa. $5.00

Prices subject to change without notice.
Available at your book dealer or write for free catalogue to Dept. SCI, Dover Publications, Inc., 180 Varick St., N.Y., N.Y. 10014. Dover publishes more than 150 books each year on science, elementary and advanced mathematics, biology, music, art, literary history, social sciences and other areas.

D1015901